More Praise for *Climate Justice*

"Addressing climate phenomena is the way to ensure justice for humanity. Mary Robinson, as UN Special Envoy on Climate Change and as UN High Commissioner for Human Rights, has been a global champion to bring justice for all. Her book inspires and guides us on what to do to protect humanity and our only world." —Ban Ki-moon, eighth UN Secretary General, member of the Elders

"Robinson's lucid, direct style works because it gives a voice to those who have taken it upon themselves to tackle earth's most pressing problems. The book's central message is a mantra worth repeating: individual local action can grow into a global idea, producing positive change." —*The Guardian*

"Sustainable development is at the heart of climate justice— protecting the planet, now and for generations to come. The stories in this book reveal the lived experience of people doing just that, adapting and strengthening their resilience in the face of climate change. They are courageous men and women whose lessons we all should heed." —Gro Harlem Brundtland

"[Robinson] uses her powerful platform to highlight the work of mostly female climate activists in frontline communities that are already reeling from the effects of climate change . . . Written in a post-Trump world, *Climate Justice* burns with urgency." —*Sierra*

"This is a book about people: farmers and activists in Africa, Asia, and the Americas, people whose livelihood is ruined by climate change and climate injustice. Yet it is also a celebration of their fight back. I was moved by Mary Robinson's account of amazing women leading the fight for their communities."
—Mo Ibrahim

"Robinson's humility and compassion resonate through her storytelling . . . [Her] stories provide a window into our own future." —*The Irish Times*

"Exceptionally informative and impressively organized and presented . . . An erudite and documented manifesto with respect to a critically important and universal humanitarian issue."
—*Midwest Book Review*

"Mary Robinson brings the power of the voice of those heavily affected by climate change—particularly women—to the center of the consciousness of decision makers to propel collective action." —Graça Machel

"Putting a human face to those on the front lines and giving them a voice, Robinson illustrates the day-to-day impacts of climate change on those around the world, making the threat more real, more pressing, and, ultimately, more frightening . . . *Climate Justice* is a compelling, easy read that should persuade people to take personal responsibility for the problem." —*Ms.*

"Robinson makes a powerful and compelling case that the climate crisis is a crisis of humanity, requiring far more than mitigation and adaptation, but a renewed sense of shared destiny. Simply put, climate action must work for the good of all, or it won't work for anyone." —Richard Branson

"Robinson puts a human face on this politically charged issue, adding to the climate change conversation. Highly recommended." —*Library Journal* (starred review)

"Robinson is uniquely qualified to write about the international fight for climate change justice . . . A surefire winner." —*Booklist*

"Provide[s] both hope and suggestions of concrete ways we might yet respond with empathy and support to those who are suffering most in this global crisis." —*The Presbyterian Outlook*

"Diversely, refreshingly human. The data is accurate, the diagnoses far from naive, and yet it manages to generate pragmatic hope: responses are possible, and efficacy is not limited to elite corridors of power. This gem of a book can be read in quick bursts or one fell swoop and is well suited for both bedside table and academic syllabi." —*America*

"Giving voice to the previously voiceless, providing seats at the table not only for the powerful who are proceeding heedlessly, but for those who have been suffering devastating consequences . . . Hopeful and optimistic . . . [Robinson] tells engaging stories of extraordinary accomplishments by ordinary people." —*Kirkus Reviews*

Climate Justice

*Hope, Resilience, and the Fight for
a Sustainable Future*

Mary Robinson
with Caitríona Palmer

BLOOMSBURY PUBLISHING
NEW YORK · LONDON · OXFORD · NEW DELHI · SYDNEY

To those whose stories of hope and
resilience inspired this book.

BLOOMSBURY PUBLISHING
Bloomsbury Publishing Inc.
1385 Broadway, New York, NY 10018, USA

BLOOMSBURY, BLOOMSBURY PUBLISHING, and the Diana logo
are trademarks of Bloomsbury Publishing Plc

First published in the United States 2018
This paperback edition published 2019

ISBN: HB: 978-1-63286-928-9; PB: 978-1-63557-592-7;
eBook: 978-1-63286-930-2

LIBRARY OF CONGRESS CATALOGING–IN–PUBLICATION DATA
IS AVAILABLE.

4 6 8 10 9 7 5 3

Typeset by Westchester Publishing Services
Printed and bound in the U.S.A. by Berryville Graphics Inc.,
Berryville, Virginia

To find out more about our authors and books visit www.bloomsbury.com
and sign up for our newsletters.

Bloomsbury books may be purchased for business or promotional use. For
information on bulk purchases please contact Macmillan Corporate and
Premium Sales Department at specialmarkets@macmillan.com.

CONTENTS

Prologue

MARRAKECH

O N   T H E   N I G H T of November 11, 2016, in a guesthouse in the ancient medina of Marrakech, I was having diffi-culty sleeping. I had arrived that evening on a flight from Paris to attend the UN annual climate change talks. One year earlier, the signing of the Paris Agreement had marked a critical turning point towards a zero-carbon, more resilient world. Now, repre-sentatives from 195 nations, including the United States, were gathering in this Moroccan city to discuss ways to implement the agreement.

After a late-night dinner and debrief with my foundation team,[1] I retired to my room, overlooking a small courtyard inset with a turquoise-green pool. There, I tossed and turned late into the night, unable to shake a feeling of apprehension. Days earlier, in an electoral result that had shocked me and the rest of the world, Donald Trump had become president-elect of the United States. By a strange coincidence, I had been declared president-elect of Ireland twenty-six years earlier to the day.

In the previous weeks leading up to Marrakech, as my small team and I worked long hours at our office in Dublin in prepa-ration, I had kept a close eye on the electoral race playing out across America. As Election Day approached, I became increasingly anxious about the prospect of a Trump victory. I was deeply concerned by Trump's anti–climate change rhetoric and his promise to pull the United States—the world's most powerful nation and the biggest carbon polluter in history—out of the Paris Agreement, which had come into force just four days before the election. One of the greatest international achievements in multilateral diplomacy, the agreement was a shining example of how the world could come together to combat an existential global threat. How would Trump's election affect the resolve of other countries in Marrakech? In my heart, I knew that the Paris Agreement was stronger than any one nation, yet I felt foreboding at the prospect of this new U.S. administration.

So much was at stake. For more than a decade, I had met those suffering the worst effects of climate change: drought-stricken farmers in Uganda, a president struggling to save his sinking South Pacific island nation, Honduran women pleading for water. They come from communities that are the least responsible for the pollution warming our planet, yet are the most affected. They are often overlooked in the abstract, jargon-filled policy discussions about how to address the problem. But their stories have made me realise that the fight against climate change is fundamentally about human rights and securing justice for those suffering from its impact—vulnerable countries and communities that are the least culpable for the problem. They must also be able to share the burdens and benefits of climate change fairly. I call it climate justice—putting people at the heart of the solution.

The next morning, I awoke in Marrakech resolute on a course of action: I would issue a statement urging the United States to stay the course and to withstand any efforts by a Trump White House to derail the Paris Agreement. Over breakfast, I discussed my plan with the director of my foundation, who voiced caution. Next, I called my husband, Nick, my mentor and great ally, who was at our home in county Mayo, Ireland. Nick listened to my proposal and then gently advised me not to make a speech or statement, suggesting that it would be counterproductive to take such a sharply critical tone. Still determined, I called my friend and close adviser Bride Rosney in Dublin. "I understand how you are feeling, Mary," Bride told me. "You need to get this out of you, but you need to do it the right way. All you need is a journalist to ask you the right question."

Later that morning, in a quiet corner away from the bustle of the climate talks, I spoke on camera with Laurie Goering from the Thomson Reuters Foundation. Fighting back emotion, I described how, as United Nations Envoy for El Niño and Climate Change, I had recently met women in the drought-stricken regions of Honduras who no longer had water. I had seen the pain on the faces of those women. And one of the women said to me, and I'll never forget, "We have no water. How do you live without water?"

Laurie moved the mic nearer and I expressed my pent-up feelings: "It would be a tragedy for the United States and the people of the United States if the U.S. becomes a kind of rogue country, the only country in the world that is somehow not going to go ahead with the Paris Agreement. The moral obligation of the United States as a big emitter, and historically an emitter that built its whole economy on fossil fuels that are now damaging the world—it's unconscionable the United States would walk away from it."

I felt lighter as the interview drew to a close. It was a relief to speak my mind and to define the moral issue at stake. I remembered something the poet Seamus Heaney wrote to me on the day that I became the United Nations High Commissioner for Human Rights: "Take hold of it boldly and duly."

By week's end, the ripple effects of the election had dissipated across the tented canopies of the climate talks, and my fears were diminished. Nation after nation—in union with civil society and business leaders—reaffirmed their commitments to the Paris Agreement. Behind closed doors the meeting buzzed with a renewed sense of urgency. Along the corridors of the climate talks, housed on the dusty outskirts of Marrakech, people moved with more energy. On the final day, forty-eight of the poorest countries made an extraordinary pledge: They would receive all their energy from renewable sources by 2050. Having some of the countries most vulnerable to climate change lead on delivering the goals of Paris was a powerful and humbling declaration. The message was clear: There was no turning back. The rest of the world would forge ahead with or without the United States.

UNDERSTANDING CLIMATE JUSTICE

O N DECEMBER 12, 2003, my thirty-third wedding anniversary, I was at a meeting in Trinity College Dublin when my cell phone rang. It was my son-in-law, Robert, breathless with news. My daughter, Tessa, had just given birth to their first child, a boy. Could I come to the hospital, Robert asked, and meet my first grandchild?

I grabbed my coat and stepped out into the brisk winter air. It was a ten-minute walk from Trinity College through the heart of Georgian Dublin to Holles Street and the national maternity hospital, where, thirty-one years earlier, I had given birth to my own first child, Tessa herself.

In the hospital ward, I embraced the exhausted but elated couple. Tessa tenderly passed me a tiny bundle and watched with delight as I peered inside. Face-to-face with my grandson, Rory, I was flooded with a rush of adrenaline, a physical sensation unlike anything I had ever felt before. In that moment, my sense of time altered and I began to think in a time span of a hundred years. I knew instinctively that I would now view Rory's life through the prism of our planet's precarious future. I made a quick mental calculation: In 2050, when Rory would be forty-seven, he would share the planet with more than nine billion people. These billions would be seeking food, water, and shelter on a planet already suffering the effects from our global dependency on fossil fuels. What would that world be like? Would we have pushed ourselves by then to the verge of extinction? The abstract data on climate change that I had skirted around for so long suddenly became deeply personal. Holding this tiny baby, I instantly felt the threat that climate change could pose to him—and thereby to all humanity. I would be long gone by 2050, but what could I do to help ensure that Rory, and every other baby born in 2003, would inherit a world fit to live in, and not one on the brink of despair?

∞

I'm humbled to admit that I am a relative latecomer to the issue of climate change. When I served as United Nations High Commissioner for Human Rights from 1997 to 2002, safe in the knowledge that the United Nations already had a dedicated climate change office, the topic rarely crossed my mind. I don't remember making a single speech relating to it. That changed in early 2003 after I had moved to New York to create my own organisation, Realizing Rights: The Ethical Globalization Initiative, to advance economic, social, and cultural rights, particularly in African countries. As high commissioner, I had watched industrialised nations emphasise the importance of civil and political rights while rarely conceding that the right to food, safe water, health, education, and decent work were equally important. With Realizing Rights, I wanted to change this dynamic, to make human rights matter in small places, and to help developing countries achieve their full economic and social potential. I wanted people in developing countries to know they had inherent human dignity and rights, and that those in power should "realise" those rights by implementing and respecting them.

But from the outset of my Realizing Rights days, as I travelled in African countries to promote the right to development, an unexpected issue kept getting in the way: climate change. No matter where I went, I kept hearing variations on the same phrase: "But things are so much worse now." Farmers in Africa described the erratic nature of their harvests, how they failed to arrive when expected, and how long months of drought would be followed by flash floods that swept away farms and villages. Across the Americas and Asia, people told stories of hurricanes that destroyed homes and hospitals and took out government services, schools, and businesses. In the past, I had seen images of stranded polar bears and the disappearance of ancient glaciers, but these anecdotal stories from the front lines of climate change suddenly began to match the scientific findings I was reading with increasing concern. It seemed that Mother Earth was trying to tell us something—that depleting the earth's resources at an ever-accelerating rate would ultimately lead to our own demise.

Climate change, I realised, was no longer a scientific abstraction but a man-made phenomenon that impacted people— primarily the most vulnerable—all over the world. While industrial nations continued to build their economies on the backs of fossil fuels, the most disadvantaged across the world were suffering most from the effects of climate change. Though these communities were the least responsible for the emissions

causing climate change, they were disproportionately affected
because of their already-vulnerable geographic locations and their
lack of climate resilience. The rising sea levels, produced by the
melting glaciers I had observed rather passively, were the cause
of water surges thousands of miles away in the low-lying islands
of the Pacific and Indian Oceans that wiped out entire villages
and livelihoods. I began to understand that climate change
was more than just the sudden violence of a hurricane or flood;
gradually changing weather patterns and rising sea levels were
slowly and steadily causing greater food shortages, pollution,
and poverty, putting decades of development advances at risk.
This injustice—that those who had done least to cause the
problem were carrying the greatest burden—made clear that to
advocate for the rights of the most vulnerable to food, safe water,
health, education, and shelter would have no effect without
our paying attention to our world's changing climate.

∞

The Industrial Revolution, which began around the middle
of the eighteenth century, started global warming by magni-
fying the release of carbon dioxide and other heat-trapping gases
into the atmosphere. Over the subsequent centuries, countries
across Europe and in America have transitioned from rural,
agrarian societies to industrial, urbanised ones, becoming rich
by using fossil fuels, especially coal and oil, to power their
economies. As wealth and consumption grew from fossil fuel
use, so too did the rate of greenhouse gases released into the
atmosphere, aided by unsustainable land use from poor agricultural practices and deforestation. Christiana Figueres, a brave
Costa Rican diplomat who led the UN body tasked with

persuading the world's 195 countries to lessen their dependence on fossil fuels, likens this process to pouring dirty, poisonous sludge into a bathtub with a partially opened drain. Carbon dioxide from burning trees and fuels flows into the world's bathtub at a faster rate than it can be drained by the atmosphere, plants, and the ocean. By the second half of the twentieth century, scientists were sounding the alarm that the bathtub was about to flood. The stock of greenhouse gases already in the atmosphere was causing a dangerous rise in global temperatures—leading to sea level rises and dramatic changes in weather patterns. Those same scientists made clear what needed to be done: We had to stop pouring filth into the tub and work collectively to drain it.

By the 1990s, some courageous and scientifically minded political leaders had awoken to the seriousness of the threat. At the Earth Summit in Rio de Janeiro in 1992, the United Nations Framework Convention on Climate Change (UNFCCC) was created to coordinate global efforts to combat climate change and deal with its consequences. This summit in Rio would lay the groundwork for the Kyoto Protocol, an initiative that encouraged rich nations, which had already benefitted from industrialisation, to reduce their greenhouse gas emissions and to do so before developing nations would join in. But the resolve shown at the historic summit in 1992 did not translate into action at the scale or pace needed, and the Kyoto Protocol did not come into force until 2005. The United States, the world's biggest emitter at that time, also failed to ratify the protocol. As is so often the case with a shared international issue, getting everyone to agree that a problem needs to be solved is far easier than getting individual countries to agree on what they themselves will do to solve it.

There is universal agreement that total global warming should be kept below 2° Celsius (3.6° Fahrenheit) or as close as possible to 1.5°C above pre-industrial levels. Two degrees of warming has traditionally been considered the threshold beyond which the effects of climate change move from treacherous to catastrophic, but most experts agree that we are already on track to exceed that. To go above 3°C or 4°C, scientists warn, will initiate a "tipping point" in our planetary system from which there will be no turning back.

As of early 2017, earth had warmed by roughly more than 1°C since 1880, the year when records were first taken on a global scale. Although this figure may seem quite low to the average person, the rise has set off alarm bells within the global scientific community, who warn that continued and unchecked levels of warming could undermine our planet's capacity to support human existence. Already, soaring temperatures around the world are setting records: the month of March 2017 represented the 627th month in a row of warmer-than-normal temperatures.[1] The five hottest recorded summers in Europe since the late-medieval ages have occurred since 2002.[2] In 2015, the Middle East and the Persian Gulf registered record temperatures as high as 73°C or 163°F.

In 2014, the Intergovernmental Panel on Climate Change (IPCC), a UN network of leading climate experts, issued a report warning that should the world remain on its present trajectory, we will hit four degrees of warming by the end of this century. Warming of more than 1.5°C above 1880 levels would lead to the loss of 90 percent or more of all coral reefs. An increase of 2°C would almost double current water shortages around the world and lead to a massive drop in wheat and maize harvests. The vicious heat waves that we experience today would become the norm, and the inundation of coastal cities like that of Houston, Texas, in August 2017 would become routine, forcing tens of millions of people to lose their homes. The sci-fi movies depicting giant storms or weather events that threaten our very existence may one day seem less fictional: a 3.6°C rise above pre-industrial levels, the IPCC warns, would precipitate an "extensive" extinction of species across the globe, rendering much of the globe uninhabitable.

Soon after Realizing Rights was formed, I was invited to chair the board of the International Institute for Environment and Development (IIED), a think tank based in London to promote sustainable development worldwide. This wonderful institution taught me how important it is to listen to grassroots voices and enable them to be heard. In 2006 I was asked to give a lecture to honour the life and work of the organisation's founder, the English economist and writer Barbara Ward, an eminent intellectual and moral leader, whose legacy was rooted in the belief that the environment and development are fundamentally linked. Human beings, she once wrote, had forgotten how to act as good guests on earth and to tread lightly on our planet as other creatures do. Several seminal events in 2006 had pushed climate change to the front pages of some leading international newspapers, and public perception of the debate appeared to be changing. Millions of people across the world had flocked to see the eye-opening film *An Inconvenient Truth*, by former U.S. vice president Al Gore. That same year, widespread heat waves, unlike anything seen before, had killed hundreds across America. In the United Kingdom, an influential report, *Stern Review: The Economics of Climate Change*, authored by Lord Nicholas Stern, had concluded—to wide international coverage—that investing now to limit climate change and to prepare for its effects would cost a fraction of the measures needed if we wait until these adverse impacts make themselves known.

Honouring Barbara Ward, I reminded those present of the words enshrined in Article 1 of the Universal Declaration of Human Rights stating our birthright: "All human beings are born free and equal in dignity and rights." Yet, when it comes to the effects of climate change, there has been nothing but chronic injustice and the corrosion of human rights. "For too long, many countries have denied the evidence, seeking to find excuses for inaction," especially the United States and Australia, which failed to live up to the clear moral obligation of signing the Kyoto Protocol. "We can no longer think about climate change as an issue where the rich give charity to the poor to help them to cope with its adverse impacts." Success would depend on a new spirit of multilateral efforts with rich countries living up to their responsibilities for contributing most to the problem. "If there is a climate change problem, it is in large part a justice problem. Our continued existence on this shared planet demands that we agree to a fairer way of sharing out the burdens and benefits of life on earth, and that in the choices we make, we remember the rights of both today's poor and tomorrow's children."

To deal with climate change we must simultaneously address the underlying injustice in our world and work to eradicate poverty, exclusion, and inequality. That injustice is embodied in the fate of the 1.3 billion people around the world who still have no access to electricity, and the 2.6 billion who still cook on open fires. If we are to properly address climate change, we must do so in tandem with improving the lives of these people, by giving them access to electricity and cookstoves through renewable energy sources, and not fossil fuels. By doing so we can deliver a wave of empowerment in one of the most profound attacks on global poverty and inequality ever to take place—by opening unprecedented opportunities for these billions of people.

Raising awareness about climate justice requires us to marry the standards of human rights with issues of sustainable development and responsibility for climate change. We need to create a "people first" platform for those on the margins suffering the worst effects of climate change, and to amplify their voices to ensure them a seat at the table in any future climate change negotiations. In the words of Archbishop Desmond Tutu of South Africa, climate justice can be a new "narrative of hope."

∞

Towards the end of 2010, I brought Realizing Rights to a planned end and moved back home to Ireland to establish my own charitable foundation on climate justice. Using the Universal Declaration of Human Rights and the UN Framework Convention on Climate Change as a scaffold, we pieced together the foundation using the principles of human rights and combining issues of sustainable development with responsibility for climate change.

In November that year, Mexico held the sixteenth Conference of the Parties, or COP—the annual climate summit meeting of the "supreme body" of the UNFCCC. A year earlier, Denmark had hosted the fifteenth COP in Copenhagen, a summit marked by disorganisation and frayed tempers when numerous small states were sidelined by the bigger powers and left—literally—outside in the Danish cold. I and many others were anxious about what lay ahead in Cancún, worried that negotiators would not be able to revive the climate talks after the disaster of the previous year. The Mexican organisers, under the leadership of Minister for Foreign Affairs Patricia Espinosa, had spent the difficult year in between reaching out to the 192 countries that would be present at Cancún, trying to smooth the water and make the conference a turning point, an opportunity to get the climate talks back on track.

It also turned out to be a serendipitous celebration of the power of women. Patricia Espinosa would become one of three consecutive women—behind Connie Hedegaard of Denmark and ahead of Maite Nkoana-Mashabane of South Africa—to preside over an international climate change conference. Playing on this synergy of a "troika" female leadership, I suggested to the Mexican government that we cohost an event for women leaders on gender and climate change at Cancún, to sound out a theory that had been percolating: that by linking women with access to power to women on the front lines of climate change at the grassroots level, we could all gather strength and create a new kind of climate activism. Patricia Espinosa immediately suggested we form a network under the leadership of herself, Connie Hedegaard, and Maite Nkoana-Mashabane.

The following year at the climate conference in Durban, South Africa, this Troika alliance would become a Troika Plus of fifty female leaders, ministers, and heads of UN agencies—most of them mothers, some grandmothers. An unusual inclusive style to our meetings developed later as Troika leaders began to bring grassroots women to climate meetings to bear witness to the effects of climate change in their communities. Around conference tables usually reserved for multilateral leaders, these grassroots women would tell their stories: their fears, their frustrations, and their efforts to gain access to the simple necessities of life that so many delegates took for granted.

The Troika Plus injected a new sense of energy into all climate justice work. All at once, it became possible for women across multiple sectors of society to lead the way in helping to build resilience, amplifying the voice of ordinary women in the negotiations on climate change. In April 2013, my foundation joined with the Irish government in organising a major climate change conference to coincide with Ireland's presidency that year of the European Union. Entitled "Hunger, Nutrition and Climate Justice," it brought more than a hundred grassroots climate change activists to Dublin. Although the conference welcomed leading dignitaries such as former U.S. vice president Al Gore and EU commissioners, the promise of hearing from grassroots participants strongly encouraged others to attend.

At the outset of the conference, a weekend induction course bolstered the confidence of the grassroots participants before they sat at roundtable discussions with ministers and others. On the second day of this induction, an alarm bell was raised when it appeared that two participants, female herders from Mongolia, were missing. The women had been seen first thing in the morning but had subsequently vanished. A search party was organised but the women were nowhere to be seen. They returned later that night, windswept but joyous, having thumbed a lift all the way from Dublin to the Cliffs of Moher, nearly three hundred kilometres away on the opposite side of the country. Having spent all of their lives on the rolling plains of landlocked Mongolia, the women had heard mystical things about the great ocean that crashed upon Ireland's western shore and thought nothing of a seven-hour round trip to take a look. "We wanted to see the sea," they offered.

The following day, these women from Mongolia hunkered down and shared stories with women from Inuit and Latin American communities, all finding common ground in their climate change experiences, personal stories, and solutions. Individual grassroots participants, emboldened by their weekend training, confidently took the floor and challenged leaders, including government ministers, on what actions they could take to better effect change. This was a much-needed dose of reality for the assembled dignitaries: Hearing firsthand the experiences of those suffering from the effects of climate change was a humbling reminder of the power and principle of participation.

Being around these grassroots participants was also a powerful reminder of how lost decision makers can become in the inaccessible jargon of "international development–speak." A participant from Zambia, who had listened intently throughout the conference, finally raised her hand at a senior-level roundtable. "I have been hearing this expression—'we need to think outside the box'—for the past three days," she said, reiterating the cliché. "It seems a little strange to me," the woman continued with bemusement. "In my community, we don't think in boxes."

That exchange would be the first of many I would have with extraordinary people who have endured the drastic effects of global warming and are striving to help their communities adapt. Their stories of resilience and hope, and their quest for climate justice, can light the path forward. These women and men from Kiribati to Uganda to Mississippi can guide us as we grapple with humanity's greatest challenge.

Constance Okollet, a self-described "climate change witness," testified at a Cape Town hearing: "We, the people at the grass roots level, are suffering the worst effects of climate change." (© Edward Echwalu)

2

LEARNING FROM
LIVED EXPERIENCE

A NOMADIC HERDSMAN FROM Kenya, Omar Jibril, tentatively approached the microphone. In a halting voice, he described how the pastures in his northeast region had shrivelled to nothing in the wake of a devastating drought, practically wiping out his herd of Boran cattle. "I had two hundred cows, but now I have only twenty left," he said. "They have all died. Imagine this: no money, no food for the animals, no food for your children."

Omar was giving testimony in October 2009 at a "climate hearing," one of seventeen special tribunals being held around the world by Oxfam to gather evidence from frontline witnesses to the effects of climate change. The collective testimony from these hearings would be delivered, two months later, to world leaders at the UN climate summit in Copenhagen to highlight the human toll behind global warming.

At the climate hearing, I sat on a panel next to Archbishop Desmond Tutu. Five farmers—four of whom were women—approached the bench to share personal stories of how climate change was affecting their lives. The first, a small-scale rooibos farmer from the Suid Bokkeveld region in South Africa, told the panellists how drought and rising temperatures had wiped out any gains from the sale of her organic rooibos tea to local and foreign markets.

Another farmer from Malawi, Caroline Malema, a mother of six, described the mass flooding that had devastated her region a year and a half earlier. "In the night, we heard a big noise coming from the river, and people were crying, 'Water, water!' In the morning, when we went to the river, we saw that everything had been swept away and the cattle had been killed." Caroline and fellow villagers had tried to replant their crops, but a severe drought that descended in the aftermath of the flood destroyed the harvest. Now some women in her village, Caroline said, had resorted to prostitution to feed their families.

The next witness, Constance Okollet, a small-scale farmer and community organiser from eastern Uganda, approached the bench with quiet dignity. A self-described "climate change witness," Constance had come to Cape Town as an emissary with the nonprofit group Climate Wise Women. "I have seen live what is happening on the ground," Constance said in a soft but deliberate voice. "Everyone across the world should under-stand what is happening: that we, the people at the grassroots level, are suffering the worst effects of climate change." Constance recounted how her tiny village had been devastated since 2000 by drought, flash flooding, and erratic seasons. "In eastern Uganda, there are no seasons anymore. Agriculture is a gamble." Once believing that God was wreaking revenge on her people for some mysterious wrongdoing, Constance now knew the real cause of the unpredictable weather: "It was not until I went to a meeting about climate change that I heard it was not God, but the rich people in the West who are doing this to us. We are asking that they stop or reduce [their emissions]."

As each farmer told his or her story, the normally ebullient archbishop's body language began to shift. By the time Constance had finished speaking, Tutu was slumped dejectedly in his chair, his expression grave. He had begun the meeting with buoyant cheerfulness, but just an hour into the proceedings he was deeply affected by what he was hearing. My thoughts turned to a childhood memory of accompanying my father, a local family doctor, as he made house calls in the west of Ireland. I used to love these expeditions, sitting behind my father as he manoeuvred his car along the narrow hedge-lined roads of county Mayo. Often, we would stop to ask directions from a farmer harvesting hay in his field or pause to allow another herding sheep to move them gently to the side of the road. Conversations would invariably turn to the current weather conditions, the farmers often complaining bitterly about how excessive rain or heat had affected their recent harvest. With that in mind, I asked the five farmers in front of us whether theirs were simply other cases of farmers—albeit from other countries and generations—complaining about the weather.

Constance looked me straight in the eye and did not waver as she delivered her retort. "This is different," she told me with quiet resolve. "*This* is outside our experience."

Constance's gentle but dignified rebuke stayed with me long after the climate hearing had wound down and the participants had left the stage. I wanted to know more about her life before the ravages of climate change had descended upon her village; to understand, in specific detail, how recent weather and events

in her Ugandan village were, in her words, "outside our experience." Her story speaks to the greatest potential threat facing
our planet.

∞

When the first drops of water began to fall on a September day
in 2007, Constance Okollet tried to ignore the rain and instead
went about her daily tasks. She swept the inside of her mud-brick
house, made the fire for breakfast, and picked vegetables from
the garden adjacent to her small compound. But as the hours
wore on and the rain came in greater torrents, Constance began
to worry. As a fourth-generation farmer, Constance instinctively
understood the abnormality of the heavy rainfall that had inun-
dated her tiny Ugandan village, Asinget, that July and August.
The rainy season in eastern Uganda generally lasted from
February through April, with the rains returning once more in
October and November. The months of June through September
marked a respite from the wet season and a time for Constance
to harvest her crops. The persistent rain that had fallen that July
and August was troubling and uncharacteristic. For nearly seven
years, Constance had noted dramatic changes in the weather—
longer rainy seasons followed by intense periods of drought—
that withered her maize, sorghum, and millet, weighting her
produce with moisture and pests, crippling her yields. The
unpredictable weather, Constance worried, was a warning that
the villagers of Asinget had done something calamitous to incur
God's displeasure.

But the rain that September day was the heaviest yet,
and as dusk approached, a flash flood descended upon the
village. Surveying the brackish water engulfing her compound,

Constance knew that she had little time left. Gathering their seven children, Constance and her husband joined the swarm of residents leaving the village, making for higher ground and the safety of her sister's house several miles away. "The floods were unlike anything we had ever seen before," Constance would later relate. "They covered our village and washed away everything. Houses sank, crops and animals were swept away, and people perished in the water."

News reports described the heavy rains falling in Uganda, and the severe floods that raged through the east and north of the country, making roads and bridges impassable. In the worst-affected areas, schools, homes, health centres, and other infrastructure were destroyed or badly damaged. With tens of thousands of people displaced from their homes, the Ugandan government and humanitarian agencies rushed to provide temporary shelter, food, drinkable water, sanitation facilities, and medicines. Meteorological experts predicted that Africa would suffer the worst consequences of our warming planet, with the likelihood of more floods, droughts, and landslides—along with diseases such as typhoid, cholera, and malaria. In 2007, the year that Constance's village flooded, twenty-two African countries experienced their worst wet seasons in decades, with devastating rains affecting over 1.5 million people.[1]

For two weeks, Constance and her family crowded into her sister's house until it was safe to return home. Although most of Constance's neighbours' mud-brick houses had crumbled in the wet conditions, her house—although badly damaged— remained intact. She immediately set about shoring up her disintegrated walls and invited her neighbours in. By nightfall, twenty-nine people were bedded down on the damp floor. "We had no food because the granaries had been destroyed," said Constance. "There was no clean water to drink, and people got cholera and diarrhoea. Because of the standing floodwater there were a lot of mosquitoes. Members of my family became ill with malaria."

With her family's vegetables and stocks of cassava, millet, and sorghum washed away in the flood, they had little or no food to eat. Reluctantly, they approached the local government for help. It was a humiliating moment for a woman who had never asked for government assistance. "Now we were beggars. They gave us a mug of beans, a half a kilo for the entire family. It wasn't enough so we decided to ask for seed instead."

With quick-maturing seeds provided by the government, Constance set about replanting her destroyed gardens. But the seeds struggled to take hold in the dry, scrappy soil, the aftermath of a severe drought that had descended upon the region following the flood. "After the floods, we had no rain for six months, not even a single drop. The topsoil was very thin after the floods, but it was further eroded by the drought. The plants, particularly the cassava, all dried up. People started to die of famine. Things were completely turned around. People began asking themselves, 'Why is this happening?' "

∞

Rural women in Uganda live a grueling existence. For Constance, that means rising at five A.M., removing the goats, sheep, and chickens from her kitchen, and sweeping the straw floor free of dung, feathers, and goat hair. She then walks to the nearest well, one kilometre away, to collect water. Back at her house she makes fire for the breakfast, feeds her family, and spends the rest of the day working in the fields. She will return to the well at least three times throughout the day and make lunch and supper over a woodstove using firewood that her children collect from the bush on the way home from school.

As hard as it may be to take in, nearly 70 percent of the food consumed around the world is produced by millions of small-holder and subsistence farmers across Asia and Africa—the vast majority women.[2] Working not just as farmers but as the bread-winners and backbones of their communities, these women inevitably bear the greatest burden of our changing climate. "The women in my community have never had time to rest. But now, with climate change, their life is even worse," said Constance. "There is less water now, so I have to go more frequently to the well. Sometimes, when the well is low, I wake up at midnight to fetch water because the line during the day is too long. Sometimes I go to my field only to find that someone has stolen my crops. I know that my neighbours must be very desperate, very hungry, if they are forced to steal. Domestic violence is on the rise. Women have to travel farther and farther to fetch water and firewood. Some men do not understand and beat their wives if they spend too much time away from their homes."

As a child growing up in Kisoko alongside her eight siblings, Constance remembers a life of simple pleasures. Back then, when the seasons were regular, her family sowed and harvested crops and had plenty of food. "As a child, I never saw a flood or knew what a drought was. The rains were punctual. They would arrive in time every year, from February to April. Year in and year out, there was food. The granaries were overflowing with millet and sorghum. There was never any sickness. Now, in my community I see malnourished children. Children with an old person's face. A child now may have only half of what it used to eat and will need to spread this portion over two or three days."

∞

From my travels around the world over the years, I have witnessed, time and time again, the remarkable role of women as agents of change. When faced with insurmountable odds, it is usually women—in the home, in the local community, at the grassroots level—who are likely to organise and make their presence felt. And so it was Constance, who, with other women in her village, decided—in the wake of the devastating floods— to take matters into their own hands and to form a group to help one another. Frustrated by the slow recovery of her village and determined to improve the lives of local women, in 2008 Constance formed the Osukuru United Women Network to help bind her community together. The group was named after the subcounty, Osukuru, where Constance lived. Beneath the giant mango tree in her dusty red compound, Constance invited her neighbours to gather each week and share their problems. While the chickens scratched in the dirt around their feet, villagers spoke of hunger and frustration, how the parched earth failed to yield any crops, how their animals were dying of thirst and disease. They talked about how the schools in the villages were empty, the children too weak to learn. Others spoke in whispers of children being married to other children, a means for desperate parents to ensure that their young might find a chance of survival in another home. Constance collected these testimonies and presented them to the local council. In turn, the council slowly responded, distributing better inputs—seeds and fertiliser—and sowing equipment. Little by little, Constance found her voice. Emboldened, she set up a credit union to encourage women to pool their savings. The group met weekly and selected members to receive small loans—enough to buy a new hoe or a sack of flour, or medicine for a sick child.

In the spring of 2009, Constance heard that an NGO called Oxfam was holding a meeting about food insecurity in the nearby town of Tororo. At the meeting, Constance told the Oxfam representatives of the terrible drought and hunger that had afflicted her village. A few days later, an Oxfam representative called to ask Constance if she could attend another meeting in the Ugandan capital, Kampala, more than one hundred miles away. At that meeting, for the first time, Constance heard the words *climate change*. "I learned that overpollution from developed countries had caused real changes to the climate," she later recalled. "I felt bad because I knew that the people in developed countries are our friends. We are the same people; we have the same blood. But these people were enjoying their life while we were suffering. I wanted to know why they were doing this to us. I wanted to know whether the people in developed countries could reduce their emissions so we could have our normal seasons back."

When Constance returned to her village from Kampala, she immediately called a meeting. Sitting under the mango tree, her neighbours assembled, Constance was excited. Finally, she knew the reason behind her community's epic struggle against the elements. "I told them about climate change, about overpollution, about bad farming practices, and how there were too many cars in cities across the world," Constance said. "I told them that climate change had come to stay, but that we should try to fix things." The villagers, confused, asked Constance whether the people in the other countries around the world who were causing the pollution would come to their village to help. Constance knew better than to promise the relief of strangers from afar. "I told them that I did not know, but that instead we should try to help ourselves."

Using information from the climate meeting in Kampala, Constance urged her neighbours to consider their own impact on the environment. Forget the people in developed countries far away who were polluting, she said. What could their own tiny community do to lessen their impact on the earth around them? How could Constance and the villagers shield their community when the next rain came? Thinking back to the meeting she had just attended in Kampala, Constance recalled a presentation she had listened to on deforestation. Constance had learned about the destruction caused by the permanent loss of forests: how soil erosion occurs when trees are no longer present to anchor the soil with their roots; how the remaining good soil is washed away by heavy rains. Constance thought of the forest near her village that was slowly disappearing as her desperate neighbours cut trees to sell for firewood. Could this be the cause of her community's misfortune? Was this why the soil could not anchor their crops and homes when the rain came? Gathering her neighbours by her side, Constance took their information again to the local council. She persuaded the council to pass a law that would authorise the planting of five new trees for every single tree cut. "Now everyone is planting trees," Constance said. "Lots of mango, avocado, and orange trees. Every Sunday,

I travel to different parishes and speak after Mass. I stand up and tell the people that climate change has come to stay, but that we can overcome it by planting trees. For those who don't want to plant, I tell them to simply think of their grandchildren."

Constance's dignified voice, first heard at the climate hearing in Cape Town, has gathered potency. Now, every year when I attend the annual UN climate meetings, I look up and smile when I hear that familiar lilt. We are two grandmothers—albeit from very different circumstances—united in common concern for our children and grandchildren. For world leaders and climate change activists, Constance is a jolting force of reality, a voice from the front lines of climate change. In a room full of presidents and prime ministers, a female farmer from Uganda is the ultimate narrative of hope: an everyday climate change witness. Constance has transformed her local activism into a powerful global voice and given an incentive to other activists. She is a constant reminder of the extraordinary power of grassroots women, those who experience the effects of global warming in a personal way and use their voices and courage to effect change.

Constance once told me that her stories about the effects of climate change in her village are like water cascading into a lake. The lake, Constance says, is unable to refuse the tributary feeding into it, just like her life as a climate witness. "Our stories are like running water going into that lake. If I continue to talk, if we continue to tell our stories, the people in power, the polluters, will realise that we are still here. They will ask themselves, 'What should we do to help these people?' We should not stop speaking. We should continue the struggle. One day they will change."

Salon owner Sharon Hanshaw and other East Biloxi residents set up Coastal Women for Change (CWC) after post-Katrina federal relief failed them. "Who knew that being a cosmetologist would prepare me for a leadership role?" (Courtesy of Carmelita J. Scott)

3

THE ACCIDENTAL ACTIVIST

BEFORE HURRICANE KATRINA hit in August 2005, women in East Biloxi, Mississippi, came to Sharon Hanshaw's salon on weekend afternoons for socialising and hairstyling. While Sharon flitted between salon chairs—trimming tresses, pleating cornrows, and shaping blowouts—her clientele swapped gossip, read magazines, and waited for their manicures to dry. For twenty-one years, Sharon's salon had served as a one-stop shop for women in this coastal city overlooking the warm waters of the Mississippi Sound. While Sharon—wearing one of her signature frosted-blond or auburn wigs—tended to their hair, the women shared stories of broken hearts, of rocky marriages, of struggles with unemployment. Sharon knew that many of her clients had missed out on the economic transformation of Biloxi following the legalisation of the garish casinos that now dotted the Mississippi coast. Many more had been hit hard by the closure of the canneries and seafood factories that had once given Biloxi the nickname Seafood Capital of the World. For those women who couldn't afford Sharon's prices, she discreetly waived her fees. As the daughter of a Baptist minister, Sharon found it natural that her salon would be a place of refuge in a community rife with racial and economic disparities.

But on the morning of August 29, 2005, the hurricane hit. When the storm made landfall on the Mississippi Gulf Coast, it brought winds of 140 mph and thirty-foot storm surges crashing through the streets of East Biloxi. The floods flattened homes along the Biloxi coast, pushing enormous waves of wreckage inland into streets, and tearing apart the predominantly black communities situated on the floodplains. Before the megastorm hit, Sharon was out of town, attending the funeral of a family friend nearly three hundred miles away in Aberdeen, northern Mississippi. From there, concerned by news reports about the impending hurricane, she sent for her children. When Sharon finally returned to Biloxi, she found swaths of the city obliterated and Highway 90 fractured, the bridges entering Biloxi buckled and warped. The city's downtown roads were unnavigable, piled high with broken cars and the splintered remnants of homes and businesses that had come apart in the storm. On Bayview Avenue, Sharon's salon, ripe with the smell of waterlogged rot and mud, was completely destroyed. Next door, the rental house where she lived was in ruins, the clapboard façade ripped apart, half the roof shorn away, and the rooms knee-deep in a putrid, coagulated mud. Sharon's red sedan—the rear bumper torn off like a candy wrapper—lay upended against the wall of a nearby government building. The car, submerged in seawater for several days, was unsalvageable. The storm waters had picked up her daughter's bed and carried it a few doors down, where it lay in someone's garden.

As Sharon picked her way through the sodden remnants of her home, she salvaged only a handful of photographs and a beloved antique table that had once belonged to her mother. At her daughter's home, she spent sixteen hours hunched over the photographs, determined to preserve some pictorial evidence of her former life. But despite her efforts, the restored photographs showed only the faintest silhouettes. "I wanted to keep them because at least we would know that we did once exist," Sharon said. The mahogany table—blistered and warped by Gulf water—at first glance seemed unrepairable. Her daughters urged her to throw it away, but Sharon, undeterred, stuck it back together again with superglue. The glue hardened in ugly, lumpy streaks, exacerbating the table's sorry state.

Hurricane Katrina—and the catastrophic failure of the levees at the Industrial Canal and Lake Pontchartrain that would submerge 80 percent of New Orleans—caused more than eighteen hundred deaths along the U.S. Southern coast and damaged or destroyed more than a million homes and businesses. It displaced more than a million people, including 175,000 African American residents in New Orleans. Following the hurricane, most of the media coverage focused on New Orleans, but the destruction was just as bad, if not worse, in other places along the Gulf Coast. In Mississippi, the storm left more than one hundred thousand people homeless, put thousands more out of work, and caused in excess of $25 billion in damage. Biloxi, situated on a low-lying peninsula and floodplain, was devastated. In East Biloxi, where Sharon lived, the storm destroyed or damaged more than five thousand homes, ravaging neighbourhoods and scattering families.

For five months, Sharon lived with her daughter in Gulfport, Mississippi, until she received a government-issued FEMA trailer back in East Biloxi. Returning to her neighbourhood, Sharon was shocked to discover that the streets still resembled a war zone. Rancid piles of fly-infested debris, festering in the swampy heat, lined street corners. Rotting refrigerators, mattresses, and piles of mouldy clothes waited for trash collections that never came. Water and electricity supplies were erratic, and functioning grocery stores scarce. Thousands of East Biloxi's residents remained homeless.

Through the diminished neighbourhood grapevine, Sharon heard angry whisperings that property developers and casino owners were dominating the recovery process, lobbying officials to rebuild damaged casinos before homes. Immediately after the storm, Mississippi governor Haley Barbour persuaded Congress to award Mississippi a federal disaster aid package of $5 billion. But despite this unprecedented sum, there were indications that Sharon and other low-income residents of Mississippi would be left behind in the recovery, that assistance would be limited to homeowners with insurance. Renters such as Sharon—in addition to residents whose homes had suffered wind damage—were deemed ineligible for aid. She was outraged but did not know where to turn. Then, in January 2006, she learned about a meeting to be held in one of the few undamaged buildings in Biloxi, a funeral home across town. The following Monday night Sharon hitchhiked across Biloxi and found about fifty women—white, black, Latino, and American Vietnamese— sitting around a large table. "The women were broken-down, they had lost family members," Sharon remembers. "No one had jobs. We had just the bare necessities. Everyone was saying, 'We've got to do something.'"

∞

For those who live in predominantly low-income or minority neighbourhoods, the chances are that you, and your family, will be more severely impacted by the effects of climate change than those living in wealthier communities. As the post-Katrina devastation in New Orleans and along the Gulf Coast showed, the storm's impacts weighed more heavily upon racial minorities and the poor. Before Katrina slammed into the Mississippi coast, the state ranked first in poverty rates in the United States and had the second-lowest state median household income.[1] The hurricane devastated a ninety-mile coast that reflected the racially segregated division of neighbourhoods that had emerged in the wake of the Civil War, when African Americans were forced to reside in undesirable swampland areas prone to frequent flooding and rife with poor sanitation and unhealthy conditions.[2] Along the Mississippi coast, "out-of-town" low-value neighbourhoods sat behind the railway line that ran from New Orleans to Mobile, Alabama, creating a racial dividing line that separated these black neighbourhoods from the beachfront homes belonging to wealthy white families.

Similar patterns exist across the rest of the United States, where new evidence suggests that major cities—Boston, New York, and Miami among others—are at risk of catastrophic flooding if sea levels continue to rise, and that low-income residents of these cities living in subsidised waterfront housing will be at greatest risk. When Hurricane Sandy slammed into New York City in October 2012, many of the wealthier residents could ride out the storm in Manhattan hotels or flee inland by car. But for New York City's lower-income residents—many of whom live in apartments on flood-prone waterfront land—the storm proved devastating.

On America's West Coast, researchers[3] have found that African Americans living in Los Angeles are almost twice as likely to die as other residents of the city during a heat wave. Less likely to have access to air-conditioning, and more likely to live in buildings constructed from materials that retain heat, these inner-city residents are more susceptible to the "heat island" effect, in which heat generated by asphalt can exacerbate temperatures. Inner-city residents are also more likely to breathe dirtier air and lack access to health insurance and proper medical care. As our planet continues to warm and smoggy conditions increase, those living in low-income parts of cities will inevitably suffer the most. The poor do not have the economic conditions that allow them to be resilient in climate hazards, nor can they often choose to leave their disadvantaged areas. Policymakers tackling climate change must acknowledge these injustices for urban and minority communities caught in the crosshairs of climate change. If not, these racial disparities and economic disadvantages will only widen an already gargantuan climate gap.

The most difficult task I had faced as UN High Commissioner for Human Rights was to be secretary-general of the World Conference against Racism in Durban, South Africa, in September 2001. There had been two previous unsuccessful World Conferences against Racism, and this time—against the odds—it succeeded. I learned deep lessons about how pervasive racism and xenophobia are in our world. I learned about intersectionality, the concept that oppressive ideologies in a society, such as racism, ageism, sexism, and homophobia, don't act independently, but are instead interrelated and continuously shaping one another. Sharon's story combined poverty, sexism, and racial discrimination, which it took a group of women, led by Sharon, to counteract.

∞

Growing up in East Biloxi during the civil rights era, Sharon watched her parents and grandparents chafe against the cruelty of the Jim Crow laws, the formal system of apartheid that dominated the American South mandating the segregation of schools, restrooms, parks, restaurants, buses, trains, and drinking fountains. Even Biloxi's beaches, interspersed with WHITES ONLY signs, were deemed off-limits to African Americans. As a child in the 1950s, Sharon struggled to comprehend why she was forbidden to step upon the white sandy beaches that hugged the shoreline near her home. Instead she and her siblings would take a bus along the picturesque coast to nearby Gulfport, where they could swim and play on a small stretch of sand in front of a Veterans Administration building marked as federal property. Between 1959 and 1963, Sharon's father, the Reverend Louis Peyton, a Baptist minister and respected civil rights leader, participated in a number of "wade-in" protests along Biloxi's beaches, as he and other black men dared to venture onto a segregated beach to soak up the sun and swim. In 1960, Biloxi's second-ever wade-in led to the worst racial riot in Mississippi history, when fighting broke out across the city for several days. Several more years would pass before Biloxi's beaches were finally integrated in 1968.

Sharon's father, who also ran the Kitty Kat restaurant on Main Street in Biloxi with his wife, Mamie, taught his children that faith and hard work could overcome racial adversity. His motto, "Pray and believe, and always believe in what you can do instead of can't do," had seen Sharon through the worst of Katrina's aftermath. Now, sitting in the gloomy dim of the Biloxi funeral home that January night, Sharon thought about her father as she listened to how residents of East Biloxi were being bypassed by the federal and city governments during the relief and recovery phase. Sharon wondered how she and others from the neighbourhood could have a voice in the recovery process. "It became obvious that the poor people were the only group that was not organised. Our needs were so great but we did not have a voice."

By the third funeral-home gathering, the women had agreed that they should form a group to advocate for lower-income residents while Biloxi was rebuilt. Within weeks, Sharon and others had secured $30,000 in seed money from the Twenty-First Century Foundation to set up a nonprofit called Coastal Women for Change (CWC). Sharon volunteered to be the secretary. "I'll take notes, I'll go to meetings," she told the other women. "I'll make sure that our community is valid, that our voices will be heard." A natural leader, Sharon soon realised she had the potential to do more, and within months she was elected executive director. "Who knew that being a cosmetologist would prepare me for a leadership role," she said, laughing. "I didn't see it coming."

In 2004, a year before the storm, the casinos along the Mississippi coastline—built on floating barges, piers, or boats to honour a Mississippi law that casinos be kept off-land—brought in more than a billion dollars in revenue, and Biloxi ranked as America's third-leading gambling city behind Las Vegas and Atlantic City. But Katrina decimated Biloxi's casinos, lifting the floating structures into the air and flinging them onto land, where they scattered in pieces like toy ships. In October 2005,

Biloxi was still in the early cleanup stages when the Mississippi legislature passed a bill to change the state's gaming laws, allowing casinos to move onto dry land as long as they were within eight hundred feet of the shore. The new bill, designed to help the rebuilding of the gaming industry, was an instant success: within days several casino owners announced their intention to return to Biloxi.

Months later, the remnants of Sharon's salon and home were bulldozed and hauled away to make room for the parking lot of the newly constructed Imperial Palace Casino, a thirty-two-story structure that towered over Bayview Avenue. After the demolition, all that remained of Sharon's former life in the gravel parking lot was the tree where Sharon's mailbox was once attached. Watching the city authorities tear down the homes of friends and neighbours along Casino Row, Sharon saw that Biloxi's recovery efforts were stacked in favour of casinos, and not the people. Although the city countered that the casinos would provide much-needed jobs for the residents of Biloxi, the argument made no sense to Sharon. "These people have no cars. They have no homes," she said. "How are they going to come to work? I'm not against jobs and economic development, but homes come first, people come first. And people didn't come first following Katrina's recovery."

Weeks later, Sharon held a community-forum meeting and invited the mayor of Biloxi, city council members, and other elected officials. Over two hundred East Biloxi residents attended and put angry questions to the mayor. Within weeks Sharon and other CWC members were given five seats on the mayor's planning commission for finance, education, transportation, land use, and affordable housing. With no cars and with public transportation still crippled by the storm, it was a challenge to attend every subcommittee meeting, but Sharon pressed on. "I was at every meeting. Hitchhiking here and there. That's all I did."

Reilly Morse, a lawyer with the Mississippi Center for Justice, an organisation that provided civil rights advice and legal aid to African American and other minority residents of Mississippi, met Sharon for the first time at that community-forum meeting. Morse, a resident of nearby Gulfport, had lost his legal practice in the storm. "[Katrina] reduced my law office to a slab, left me nothing to salvage . . . and nothing to do but take bankruptcy." In the aftermath of the storm, Morse moved to Biloxi to open a Disaster Recovery Office, helping remaining residents to get emergency shelter and temporary housing and greater access to disaster-recovery grants. Despite his personal losses, Morse counted himself lucky. His home and family had survived Katrina, and he knew that many of his new clients in Biloxi were not as fortunate.

Watching Sharon navigate the blurred lines of the recovery process, Morse was impressed by her ability to cross racial and ethnic lines. "Sharon had an extraordinary skill in bringing women together across different lines. I don't think I had seen such an alliance like Coastal Women for Change emerge in Biloxi prior to that," he said. "It made a powerful impression on local leaders that they suddenly had to face women across many differences."

Forging an alliance with Morse and other local organisations known as the Steps Coalition, Sharon spent the summer of 2006 knocking on trailer doors to learn more about the needs of East Biloxi's remaining residents. By uniting under one alliance, multiple organisations and activists could connect their campaigns while preserving their own autonomy and priorities. For Morse and other activists, Sharon was a powerful collaborator despite her obvious vulnerability. "She had a kind of plainspokenness and bluntness that you see in some civil rights leaders going back many decades," Morse said. "She was an authentic voice for people who had been disenfranchised and discredited. She was speaking into that void. That was what was so powerful to me."

That summer, Sharon found elderly residents too afraid to come to the door because of rising crime levels, and young mothers unable to work for want of child care. Residents complained of fraud, price gouging, and excessive rents that were forcing some women to take up dangerous relationships that they had fled before the storm. Sharon asked the local police to increase daily and nightly patrols and surveillance and started a small in-home childcare program for working mothers. With a tiny pool of money, Sharon and CWC established a home-improvement fund, encouraging low-income families to apply for $500 grants for housing repairs. CWC created a plan to urge the development of public housing projects across the city away from flood areas. As the first-year anniversary of Katrina and a new hurricane season approached, Sharon compiled a data-base of East Biloxi's senior residents, held hurricane-awareness workshops, and helped residents prepare emergency kits to be ready for evacuation when the next storm came. "Now they have their driver's license, birth certificate, insurance paperwork, flashlight, food, water, and prepaid phone in a plastic container by their front door," Sharon said. "When it's time to go, you don't have any questions." Having survived Katrina, Sharon had little faith that city officials would come to her community's aid the next time around. "Poor communities suffer the worst," she said. "They have no ability to get out."

In the aftermath of the storm, Governor Barbour requested that Washington waive the requirement that any percentage of the $5 billion recovery package given to Mississippi be spent to benefit lower-income households. By November 2007, over two years since the storm, the state had spent $1.7 billion compensating middle- and upper-income homeowners and big businesses, and only $167 million on programs dedicated to helping the poor.[4] The message to Sharon and the CWC women was clear: low-income families would be at the back of the line—not the front—for federal aid. "In East Biloxi, you look north, east, and south and you see large investments, casinos and glittering signs," said Reilly Morse, describing the city a few years into the hurricane recovery. "But when your eyes come down to the pavement level, the roads are damaged, the storefronts closed, and there are no businesses. These people are stuck. If you are at the bottom of the economical realm, you are imprisoned in your own neighbourhood."

That same year, Governor Barbour announced that he would divert almost $600 million slated for the construction of moderately priced housing and rental units for a commercial port expansion project. Though the port expansion would inevitably increase jobs, historically—just as with Biloxi's casinos—few port jobs had gone to low-income residents. Once again, as part of the Steps Coalition, Sharon took out her clipboard and went knocking on trailer doors. Within weeks she had gathered more than two thousand signatures. Reilly Morse says those signatures were crucial in a lawsuit that the Mississippi Center for Justice was preparing against the U.S. Department of Housing and

Urban Development. "Gathering that many signatures is no small thing in a community where the cars are drowned, where the people are displaced," he said. "Sharon built this up to a critical mass to enable us to turn this situation around."

In 2008, Sharon was asked by Oxfam America to become a spokesperson on the links between Hurricane Katrina and climate change. Sharon thought the Oxfam representatives were crazy to consider a former cosmetologist from East Biloxi, Mississippi, and—in her usual direct way—told them so. "I didn't know anything about climate change. I know we have hurricanes, tornados, and floods, but I'm not an environmentalist. Shouldn't they get someone else?" Oxfam said no. In November 2009, as part of the U.S. Human Rights Network, Sharon travelled to Geneva to testify on behalf of displaced persons about the troubled recovery process after Katrina. A month later, as an Oxfam representative, she travelled to New York to attend a UN climate summit, where she provided testimony as America's first climate change witness. Soon afterwards, as a member of the Climate Wise Women's group, she flew to Copenhagen to attend the annual UN climate summit. As a climate change witness from an industrialised nation, Sharon felt insecure as she listened to people from Africa, Bangladesh, and the Pacific islands tell stories of survival from droughts, typhoons, and tsunamis. But the more testimony Sharon heard, the more commonalities she found: "Poor is poor, in any language. Just like my community in East Biloxi, these were people on the outside, on the other side of the track. I taught something to these people, and they taught me."

In Copenhagen, Sharon shared panels alongside fellow Climate Wise Women member Constance Okollet. Although coming from worlds apart, Sharon and Constance felt an instant connection. Constance's experience during the drought in Tororo reminded Sharon of the traumatic weeks after Hurricane Katrina when she and other East Biloxi residents had walked for miles to find bottled water. When Constance told Sharon about the humiliation she felt lining up for government rations in the wake of the flood that devastated her community, Sharon confessed to the shame of accepting food-assistance cards the U.S. government had distributed after Katrina. The marginalisation of the Tororo community in Uganda felt eerily familiar to what Sharon had felt all her life. "In coastal Mississippi, we're facing erosion and rising sea levels, and often it is poor, black communities that have been relegated to the most flood-prone areas. Across the globe, it is poor communities of colour where the recovery and relief efforts either fail to reach or make our survival harder."

By the end of the Copenhagen summit, Sharon had been nicknamed Mississippi Girl by Constance and had sat long into the night exchanging ideas with Constance and other climate witnesses about how to help their respective communities battle climate change. "These women got me to think about climate change, about international issues, about fair trade," said Sharon. "They gave me all these ideas. Connecting with women who were facing similar issues across the globe, and standing up and working for solutions, was inspiring. It is women who bear the brunt of climate change."

Back in East Biloxi, Sharon began to plant a community garden, an idea that had blossomed in Copenhagen following a conversation about obesity with a climate witness from the Pacific islands. By growing fresh vegetables and making them available to the wider community, Sharon hoped to tackle both obesity and food shortages while engaging East Biloxi residents in a fun activity. Alongside volunteers from AmeriCorps and members of Coastal Women for Change, Sharon spent three months clearing a large patch of trash-strewn land in East Biloxi, where they planted okra, tomatoes, peppers, corn, and beans. Sharon asked men from the community who had recently been released from jail to water and tend to the garden daily. When the first pea shoots emerged, Sharon was so excited that she screamed and ran around in circles. "To cook something that you grew is a feeling of accomplishment I have never felt before," she said. "The women from the developing world that I met at Copenhagen pushed me to think globally and helped me to act locally."

In 2011, Sharon confessed that her role as a climate change advocate was taking a toll on her life: "I feel so worn-out, mentally. I fight so hard all the time. Are we going to make it? Yes. But not without ageing and stress." By then, Sharon was living in a new home in East Biloxi, a small two-story house with calming colours and lots of skylights. In the corner of the living room sat the mahogany table salvaged from the storm that she had finally restored to its original grandeur and upon which she now ate her meals. On the walls were the framed family photographs that she had so lovingly brought back to life in those chaotic first weeks after the storm. But having survived Katrina and its terrible aftermath, on November 8, 2015, Sharon suffered the cruellest blow: a catastrophic stroke that has robbed her of the ability to speak. That vibrant Mississippi voice honed among the ruins of Katrina has now been cruelly taken away. But in the halls of Congress, in the fine print of the Paris

Agreement, and on the newly paved streets of East Biloxi, Sharon's legacy lives on.

"[Katrina] took a person," said Reilly Morse, "whose life had not prepared her to become a local, regional, or national figure, much less to appear on an international stage and speak about a broad issue like climate change. But by being authentic, by staying honest to the experience, Sharon opened the world's eyes. You don't know what your life prepares you to do until it happens. Sharon's experience shows us that no matter where you come from, as long as you have faith in yourself, you have the opportunity to make a major mark."

Patricia Cochran helps communities across Alaska and the Arctic deal with the ravages of climate change. "It has taken science a very long time to catch up to what our communities have been saying for decades." (© Alaska Native Science Commission)

4

VANISHING LANGUAGE, VANISHING LANDS

F OR MORE THAN two thousand years, the Yupik people have hunted and fished on the icy wilds of Alaska's western coast, digging holes through the frozen sea to catch salmon and stickleback and communicating to one another in an ancient lexicon that includes dozens of ways to describe ice. Passed down from generation to generation, this linguistic adaptation has helped the Yupik to navigate safely as hunters, using specific terminology to describe the ice's thickness and reliability. But with the advance of climate change, common Yupik words such as *tagneghneq*—used to describe dark, dense ice—are becoming obsolete as Alaska's melting permafrost turns the once solid land-scape into a mushy, sodden waste.

Recent scientific data confirms that the Arctic—the world's air-conditioning system—is warming twice as fast as any other place on the planet, with the average winter temperature having risen 6.3°C over the past fifty years. Alaska's soaring tempera-tures are caused by a perfect storm of confluence. When solar radiation hits snow and ice, most of it is reflected back into space. But as warming global temperatures encourage ice to melt, the exposed land absorbs the radiation, prompting yet more ice to melt. Now the people of Alaska—85 percent of whom live along the coast—are among the first Americans to feel the effects of climate change as the ground beneath them melts and gives way. This year, thirty-one native communities across Alaska are contemplating their imminent destruction because of the dependence on fossil fuel emissions of their fellow citizens farther south. These shrinking communities are confronting an impossible choice: to find tens of millions of dollars to uproot their centuries-old traditions and move their homes—and the bones of their ancestors—to higher ground; or to stay and use their limited resources to build a bulwark against the sea. Daunted by the cost of the latter, and with little in federal assistance, many are choosing to leave. In August 2016, the residents of Shishmaref, an Inupiat community north of the Bering Strait, voted to relocate their entire community from their barrier island, which has for decades been disappearing into the sea. Some experts warn that many coastal Alaskan villages will be completely uninhabitable by 2050, the year when my eldest grandson, Rory, and his burdened generation may be forced to reckon with the challenge of housing tens of millions of climate refugees.

Life in Alaska is defined by the cold, by the land, and by the people's relationship to the sea. To fish and to hunt is to live and breathe, and the rapid melting of the ice is causing many indigenous Alaskans to question their cultural identity. Nobody knows this crisis more viscerally than Patricia Cochran, who has for thirty years been working with communities across Alaska and the Arctic to help them deal with the ravages of climate change. Patricia is executive director of the Alaska Native Science Commission, but she is also a native Alaskan and Inupiat, born and raised in the coastal town of Nome, formerly a mining centre. Patricia grew up in a traditional Inupiat home, setting out across the tundra for fish camp every year and scrambling along the rocky coast with her siblings in the late-summer months, foraging for cranberries, nagoonberries, and herbs. "It has taken science a very long time to catch up to what our communities have been saying for decades," Patricia observes. "For at least the last forty or fifty years, our communities have noticed the subtlest of changes because they live very closely tied to the land. They see the changes happening in the environment around them. We were seeing the signs of climate change long before researchers and other scientists even started using those words."

As a scientist, Patricia is well aware that the changes she and her community have been observing are not caused by the Inupiat community. They are at the mercy of industrial policy based on fossil fuels in the rest of the United States. In 2015, Patricia participated in Tina Brown's Women in the World Summit in New York. Sitting beneath the bright stage lights, Patricia was cosseted in her native Inupiat dress hemmed with white fur and woven with intricate bright flowers. When asked to speak about the effects of climate change in Alaska, Patricia politely demurred and asked first that she be able to perform some protocols. This was an unusual break in the closely managed event. "In our community, it is really important to follow those protocols," Patricia said in her poised and gentle voice, "to remember where we come from, and why we are the people that we are." Invoking a centuries-old Inupiat tradition, Patricia then paid homage to the elders in the audience, our collective ancestors, and to those indigenous people on whose

Manhattan homelands we stood. This unexpected and humbling meditation shifted the energy and caused a spontaneous ripple of applause in the room. Her Inupiat ancestors petitioned, she then got down to business. "In the Arctic we have been facing these issues for many, many years. Climate change is more than just a discussion for us. It is a reality. It is something that we live with and face every single day and have for decades."

∞

As a child growing up in Nome, Patricia remembers the snow lying thick on the ground most of the year, and the sea—a single block of ice—stretching far towards the horizon late into the summer months. The winters were long and brutal, the summers exceedingly brief. But over time, the winters began to arrive later and to rush prematurely into spring. Now, when Patricia returns to visit her childhood home, the vast expanse of ice is gone, replaced instead by an open glistening sea. "We have had to build a seawall in Nome because the sea ice that used to sit in front of our villages is no longer there," she said. "That ice used to keep us safe. We have had so much rain that our fish will not dry on our fish racks. We have had such warm weather throughout the summer that berries have ripened twice in the season. Most worrying, the changing ice conditions have caused extreme erosion, flooding, and permafrost degradation across the entire community."

Permafrost, the permanently frozen sublayer of soil that has anchored Alaska for thousands of years, provides a foundation for homes, schools, and roads, and keeps the rising sea at bay. But mounting temperatures throughout the Arctic are causing this prehistoric underpinning to melt, turning the soil soggy, and releasing yet more carbon dioxide into the air. As the cycle continues, and the warming earth buckles and bends, the houses of Alaska's indigenous people sink and topple into the sea. As the dwindling permafrost exposes the soil, and the offshore ice that normally buffers the villages from storms decreases, the sea advances, eating away at the land. In the late summer, increasingly fierce storms, the results of climatic shifts, batter the coast, eroding the topsoil until it crumbles into the sea. In some places, villages have lost up to one hundred feet of land per year.

The changing weather patterns and disappearing ice consume not just the land; ice conditions across Alaska are shifting so rapidly that even Alaska's best hunters and most experienced sailors can no longer adequately predict the weather, wind, or hunting conditions. "We are losing people who are venturing out on snowmobiles to areas where the ice is thin," Patricia cautions. "There is truthfully not one of us who hasn't had an uncle or a friend who has gone out fishing and never returned home again." This rising death toll has caused some hunters to abandon their snowmobiles and return to more traditional ways of life to help them better navigate the wild. "Many have gone back to using dog teams," said Patricia. "The dogs are very smart and will not venture out onto thin ice. A snow machine will not do that for you. My young nephews are now using these traditional ways that have kept us alive for so many years."

Combining scientific expertise with her innate traditional knowledge, Patricia works to help communities across Alaska that are relocating, or those who have decided to stay. The latter are designated "protect in place," and Patricia helps these dogged communities stem the tide against the encroaching sea, knowing in many instances that their villages are already doomed.

Working with agencies such as the Army Corps of Engineers, she helps villagers with complicated applications for meagre federal funds to build rock revetments, sandbagging, and metal grading along the coast. "But in some instances, the first storm has taken out these revetments," she said. Navigating the federal bureaucracy can prove daunting. To receive funds for protection, villagers must first prove how much land has been lost, a costly and time-consuming measure. "Everybody wants to have facts and figures," said Patricia. "It's not enough to say that we are losing our community."

For years, the tiny city of Shishmaref, located on an island five miles from the Alaskan mainland, has been steadily yielding its shores—and buildings—to the frigid sea. When the Shishmaref residents voted in August 2016 to leave their land, it was estimated that close to $200 million would be needed to relocate homes and infrastructure to the new site and to build new roads, utilities, schools, and a barge landing. It is a staggering amount for a community of just six hundred residents, against which the state has offered merely $8 million. "It costs millions of dollars to move communities of just a few dozen people," said Patricia. "The sad news is that there isn't really anyone out there willing to write a check to help." A recent study by the National Academy of Sciences[1]—released in the final weeks of the Obama administration—estimated that repairing Alaska's roads, buildings, and utilities because of the damage caused by climate change will cost billions of dollars through the end of the century.

Those villagers who do make the wrenching decision to move can find it daunting and traumatic. Newtok, a village of about

350 people on the southwest coast of Alaska, has been sliding towards the Ninglick River for years by up to seventy feet a year. In 1996, the village took the pioneering decision to trade their coastal land for a more secure swath on an island nine miles south, at an estimated cost of $130 million. While several houses on the new island have already been built—stilted houses on high ground overlooking the sea—the move is painstaking and may take years to complete. While Newtok residents and elders plan their new homes and infrastructure, they still need to live where they are, maintain their daily rhythms, keep their children at school, and continue their ancient way of life hunting for moose, seals, and fish. But government agencies are not willing to spend money on a condemned village while a new one is being built, and residents are forced to live in their crumbling homes while they wait to relocate several years down the road. The village has already lost its sewage lagoon and landfill and expects to lose its drinking water source in the next year. The village has no roads, and the boardwalks that connect the stilted homes are rotting and falling apart. Some residents fear that their centuries-old culture and identity will suffer if they move. "For communities who have been there for thousands of years, it's a difficult decision to leave everything," said Patricia. "It's not only the physical exhaustion, but the mental exhaustion and trauma that comes along with all of those things."

While the Trump administration in Washington works to dismantle the climate action policies of President Obama, Patricia is redoubling her efforts on what she and her organisation can do to help indigenous Alaskans with community-based initiatives, research, and action. She frames climate change as a human rights issue, expanding the dialogue beyond emissions and mitigation to incorporate the language of justice and humanity. She focuses her efforts on working with federal agencies to provide more accurate and more targeted weather predictions so coastal communities can be better prepared for the next storm. She is helping communities build a network of local observers so that firsthand information about ice conditions and the weather can be shared with one another.

As a self-professed EIT, or "elder in training," Patricia encourages young people to take part in her climate-justice journey, so that they too can learn the tools to live a sustainable life in their native communities. "I see that as my most important responsibility and honour," she says, "to pass on that information and knowledge to the young people who must live with the disastrous situation that we have left them in." Across the one- or two-room schools that dot the vast Alaskan coastline, new climate programs are being introduced to teach young children the myriad ways to talk about the weather—and to describe snow and ice—in their native languages. It is a way to keep endangered words such as *tagneghneq* alive, and to help these children navigate a safer future. "If nobody else is going to help us, what can we do to help ourselves?"

When Patricia is not travelling across Alaskan airspace to meet with native communities in remote areas, she travels to the Lower 48—how Alaskans refer to the rest of the U.S. mainland—to lecture about climate change in her community. It is uphill work. "I spend a lot of time convincing other people of the issues that we see here in the Arctic," she says. "To many of the audiences that I speak to in the United States, this issue is still something new. It never ceases to amaze me that there are people who don't believe in climate change. I try to tell stories so that these audiences will have a different frame of reference and understand that these are other human beings. I'm trying to get them to see what is on their horizon."

Alaska's fight against the effects of climate change has a wider lesson for the rest of the United States. The searing heat that made 2016 the hottest year ever recorded continued unabated into 2017, causing unprecedented temperature rises across the world, polar melting, and rapidly rising seas. If America cannot effectively marshal enough resources to help tiny communities along Alaska's submerging coast to relocate, what will happen when the truly cataclysmic effects of climate change come? Climate change experts predict that by 2100, sea levels could rise high enough to submerge 12.5 percent of Florida's homes. In a paper published in 2017, U.S. researchers outlined how, between 2011 and 2015, the sea level along America's southeastern coastline rose six times faster than the long-term rate of global increase.[2] In 2017, Hurricane Harvey devastated the city of Houston, but other cities in the United States—Boston, New York, Atlantic City, Tampa, Miami—are vulnerable to the fiercer storms and sea level rise fuelled by climate change. FEMA, the federal agency that deals with disaster relief, is pushing communities to deal with climate change, but the federal government itself has no relocation plans, and there is no political will or funding to help communities such as Newtok or Shishmaref move away from disaster zones. "What we are seeing here in the Arctic we know is a harbinger of what is to come for the rest of the world, and for the rest of the United States,"

Patricia says. "I want to convince those coastal communities in Florida, New York, and California that they should be concerned, because climate projections show that those communities are going to be facing the same kind of issues that we are dealing with here now. To me it is just logical that you would want to have as much information as you can about what you can safely predict in the future."

While she works to help Alaska's indigenous people battle what sometimes feels like a losing fight against climate change, Patricia takes inspiration from one of her own elders, her beloved mother, who passed away some years ago at the age of ninety-six. Patricia's mother, as a young child, watched as a flu epidemic wiped out her entire family, except for her father. Bereft and traumatised, Patricia's mother was removed from her village when she was eight and sent to a boarding school, where she would remain until she was eighteen. "She lost her language. She lost her culture," Patricia remembers. "Despite this, she fought the rest of her life to make sure that her kids had what it would take for us to survive."

Patricia remembers her mother, despite this early heartache, as an eternal optimist, an indomitable spirit who taught each of her eight children to employ resilience and grit in the face of adversity. When Patricia travels across the United States and abroad, she closes her talks by telling stories about her mother, who would be working right alongside her to battle the effects of climate change. "She was a remarkable and adaptive woman. Despite all that is happening to our community now, she would be leading the way."

Keeping her feisty mother in mind gives Patricia the focus that she needs. And it helps imbue her message with hope.

"When you come from an experience like my mother's, it really makes me understand that we can deal with anything," she says. "We have always been resilient, adaptive, creative, amazing people—which has helped see us through the darkest of times in the past. That resilience, that spirit, will help us in the times yet to come."

Hindou Oumarou Ibrahim's family farms cattle in the Republic of Chad. "During the rainy season, we would milk twice, morning and evening. Now we milk only once every two days, and get one cup if we are lucky." (© The Women's Forum/Sipa Press)

Jannie Staffansson belongs to one of the few indigenous groups in the European Union, the Saami people. "As a reindeer herder, you need to be able to predict the weather to stay safe." (© Annick Ramp)

5

A SEAT AT THE TABLE

As a child growing up in the late 1980s in the Sahel region of the Republic of Chad, Hindou Oumarou Ibrahim loved to help her grandmother milk the family's red longhorn cattle. Standing beneath the heaving bellies of the majestic beasts, Hindou would hold a tin can while her grandmother expertly squirted the warm liquid inside. Hindou's grandmother would sell the milk at market but keep vast reserves of thick cream for Hindou and her four siblings. There was always plenty to go around. "During my childhood, a cow would yield two litres a day," Hindou remembers. "During the rainy season, we would milk twice, morning and evening. Now we milk only once every two days and get one cup if we are lucky. One cup every two days is not enough to feed the children, or to sell to buy cereals to eat."

The effects of climate change are drastically altering the ancient traditions of the Peule-M'bororo, a group of an estimated 250,000 nomads who are one of the most sacred pastoralist communities in Chad. M'bororo herders travel great distances across the Sahel to graze their livestock, often crossing borders into Cameroon, Niger, Nigeria, and the Central African

Republic to find the best pastures or water supplies. Some members of the M'bororo are completely nomadic, travelling across the entire Sahel, while others, especially those who herd cattle, remain in Chad and Cameroon. As ancient nomads, the M'bororo have relied on centuries of finely honed traditional knowledge to help cope with seasonal weather patterns. By observing the positioning of the stars or changes in the direction of the wind from east to west, the M'bororo can predict when the rains will fall. Finely attuned to the elements, they know that when certain birds build their nests high in the trees, the next rainy season will bring flooding, or that the proliferation of a certain insect is a sure sign of rain—even if the skies are abundantly clear. This traditional knowledge has been a soothing constant in the community's life and preserved their ancient ways.

But now the reliability of the M'bororo's weather predictions has been undermined by increasingly erratic climatic patterns, desertification, and drought. Lake Chad, once one of the largest bodies of water in Africa, has shrunk in the past fifty years from almost 10,000 square miles to just 965—a combination of the effects of inefficient damming, irrigation, and climate change. The vibrant ecological system of the lake that once yielded fish, fed millions of acres of agricultural fields, and provided grasslands for grazing has all but vanished. Cattle are producing less milk and are dying of thirst. Large swathes of lush green land that for generations supported nomadic cattle grazing are now dry and brittle. Pasture plants that provided nutritious grazing for cattle have disappeared, replaced by new varieties that make the animals sick.

The waterholes bordering Lake Chad that Hindou loved to play in as a child have vanished, forcing the M'bororo from their traditional nomadic routes to new territories. "In the past when we would move from place to place, we would stay one week minimum and up to a month maximum," said Hindou. "Now, the M'bororo stay just for three days because there is no water and no pasture. The waterholes that I swam in as a child have dried up and disappeared. The community must move constantly, and in different directions, to look for food. They travel through corridors of death." Because the M'bororo must compete for grazing land with other pastoralists and subsistence farmers, frequent—and sometimes fatal—conflicts erupt over scarce fertile land. Often when the M'bororo return to land that they previously cultivated, it has been taken over by another community.

As in Constance Okollet's village in Uganda, the new constraints imposed by the changing climate impact the M'bororo women the most, as they are forced to travel farther into the desert to collect water and food. More than seven million people around Lake Chad are now suffering from severe hunger, with half a million children acutely malnourished. With dwindling milk supplies, the women must supplement their families' diets with unfamiliar foods such as corn, maize, and rice. "The entire food system has changed, and there is a lot of sickness," said Hindou. "Because of the lack of access to water, the M'bororo now drink the same water as the cattle. It isn't clean." Many M'bororo herders from Hindou's community have abandoned their traditional way of life to become seminomadic or settled, their livestock wiped out by disease or thirst, or sold to keep their families alive.

Confusion and frustration reign among the elders as they navigate a world where their traditional knowledge and seasonal predictions are no longer in sync with the Sahel's unpredictable weather. Frightened that their generations-old relationship with the earth is under threat, and that their trustworthiness within the community is fragile, the elders invoke higher powers. "They believe that perhaps we are bad people, and so we must pray and make sacrifice," said Hindou. "They do not know the origins of climate change. They do not see any solution coming to them. They just believe in God, and that God will bring a solution."

∞

Just as Hindou's early years hummed to the rhythms of her family's nomadic life, over 3,500 miles away, in the depths of the Nordic wilderness, Jannie Staffansson's childhood was spent bustling to the needs of her family's vast reindeer herd. Jannie belongs to the Saami people—one of the few indigenous groups in the European Union—who have for centuries lived in a 150,000-square-mile area covering the northernmost reaches of Norway, Sweden, Finland, and the Kola Peninsula in Russia. Approximately one hundred thousand Saami are scattered across this isolated tundra, with an estimated twenty thousand living in Sweden alone. An ancient nomadic people who have herded reindeer for generations, the Swedish Saami have often been the victims of systemic discrimination and colonisation within their own country. In the 1930s many Saami children were forced to attend state-run boarding schools and forbidden from speaking their own language. Nowadays, oil, gas, and wind-power projects encroach upon Saami villages and threaten the Saami herd stock and the villagers' way of life. Although vast numbers of Saami have left the tundra and moved to Swedish cities farther south, many still experience prejudice for their indigenous origins.

Growing up on the Sápmi—the Saami area in Sweden where reindeer outnumber humans three to one—Jannie and her siblings spent any free time outside of school helping their father with his reindeer herd. Following the Saami calendar, which has eight seasons based on the life cycles of the reindeer—as well as the traditional seasons, the Saami observe spring-winter, spring-summer, summer-autumn, and autumn-winter—Jannie, swathed in brightly patterned traditional woollen clothing and thick reindeer-hide boots, learned to cope with long days spent outside in subzero temperatures. In the summer, during marking season, Jannie's father taught his children how to lasso reindeer calves and, with a sharp knife, expertly nick notches in the animals' ears. In the winter, Jannie's father was absent for many weeks as he moved the herd great distances to their winter pasture, where they foraged for lichen on the high branches of coniferous forests, ever watchful of predators such as wolves, bears, and lynx. "I was brought up believing that the reindeer came first," Jannie said. "My father was not present for Christmas, birthdays, or school events because reindeer herding was more than a full-time job. We were at peace with that. Tending to the reindeer was the most important thing."

Over the past thirty years—given their proximity to the Arctic, which is warming at a rate twice that of the global average—the Saami have noticed major changes in the natural environment and weather. As a child, Jannie remembers listening to her parents and village elders talk about the changing seasons—how the autumns were longer and wetter, the winters warmer, and the spring unpredictably early. In a rich lexicon tied closely to nature—the Saami have more than three hundred ways of saying *snow*—Jannie's father described the troubling impact of the changing weather on his herd: how temperature swings caused snow to melt and then to freeze over again, locking the nutritious lichen that reindeer eat under a hard sheet of ice. "These ice crusts made it impossible for the reindeer to smell the lichen underneath," said Jannie. "The reindeer would just keep on walking, spending a lot of energy trying to smell and locate the food, not knowing that it was right underneath." The warmer winter temperatures also caused the lakes and rivers to freeze over much later than usual, affecting the herd's migration path to the winter pastures. Knowing of entire herds lost from falling through flimsy ice, Jannie's father would be forced to make significant detours, causing undue stress on the reindeer. "As a reindeer herder, you need to be able to predict the weather to stay safe," said Jannie. "More and more reindeer and people are falling through the ice."

Watching her father adapt to the changing environment fascinated Jannie. When she was about ten years old, she remembers encouraging him to share his concerns about the changing seasons—and its impact on the reindeer—with the Swedish authorities. "They will never listen or believe us," Jannie's father, who has a high school education, gently told her. "For people in society to believe you, you need to have a higher education, and we don't have that." Her father's words proved seminal for Jannie, who decided right then that she would pursue a higher education so that those in authority would engage and listen to her. "I understand that the people who had all of the vital knowledge, the Saami, my people, were not heard because they were not well enough educated in the Western system. They had no voice, even though they had critical information."

True to her word, Jannie graduated from the University of Gothenburg with a bachelor's degree in environmental chemistry. She is currently working towards a master's degree in organic chemistry. Jannie's education, combined with the traditional knowledge passed down by her father, has given her the confidence to advocate on behalf of the Saami people, and to seek solutions to the environmental challenges that they face. "My degree has allowed me to enter rooms I would never have dared enter or had access to before. Suddenly, people take me seriously." But there are challenges too, remnants of the old racial prejudices against the Saami that Jannie's grandmother—who as a child was barred from using her native tongue—once faced. "Some people have asked me whether I am really of Saami origin because of my chemistry degree," Jannie remarked. "One person asked me, 'Are you sure you don't have any Swedish genes? Because the Saami really are not that smart.'"

Jannie now works on scientific and environmental issues for the Saami Council, an umbrella organisation that promotes the rights of the Saami people across Norway, Sweden, Finland, and Russia. In 2015, she was elected to a working group with the Arctic Council and played a key role in representing the group at COP21, the UN climate summit in Paris in December 2015. As a member of these bodies, Jannie has become a powerful voice for the rights of indigenous people everywhere and about the life-threatening challenges that they face. Speaking on behalf of the Saami Council, Jannie travels the world, describing the day-to-day battle that her father and his herd endure against the unpredictable weather, and the solutions that his traditional knowledge can bring. She warns against the challenges imposed by some renewable energy projects—particularly the giant windmill parks that threaten the safety and migration of the reindeer herds—and the frustration that the Saami people feel when their pleas to maintain their land rights fall on deaf ears.

In April 2017, in Brussels, Jannie and I sat together on a panel in the European Parliament organised by the Committee on Women's Rights and Gender Equality. Jannie praised the European Parliament for holding a discussion on gender and climate justice. Then she paused, looked around the hall, and said in a challenging voice, "What took you so long?" There was a moment's silence and then prolonged applause. Two years earlier, in the climax to the announcement of the Paris Agreement, Jannie had delivered an emotional rebuke when she learned that language ensuring respect for the rights of indigenous peoples was to be struck from the final draft of the landmark deal. Jannie was just one of hundreds of indigenous people who had come to Paris to fight for the survival rights of their people. She had watched with disgust while the more powerful nations had fought to defend their interests and wealth, ignoring what she felt was the only issue at hand—the fight between life and death that the Saami and other peoples on the front lines of climate change feel.

Just prior to Paris, Jannie's great-aunt had fallen through a patch of unreliable ice in the Sápmi and, despite desperate search efforts, had never been seen again. "My friends, this is the face of an angry Saami," Jannie said, her voice shaking. "We are the persons who are dying. My friends, my family, they are the ones who go through water, they are the ones who are killed in avalanches . . . How can the purpose of this negotiation not be people? How can our voices be silenced multiple times, and then again?" Her voice, and that of many others, secured a provision in paragraph 135 of the agreement that envisaged a Platform for Indigenous Peoples. That at least would be a basis for future action.

∞

Following her parents' divorce, a shocking development in the close-knit patriarchal M'bororo community, Hindou—the third of five brothers and sisters—moved with her mother and family to the Chadian capital, N'Djamena, when she was six years old. While her father would return occasionally to the nomadic life, he gave Hindou permission to attend school, a rare opportunity for an M'bororo girl. "That opportunity to receive an education was the beginning of the climate change life that I now live," said Hindou. But her father's courage in allowing Hindou to attend school ultimately caused the M'bororo community to cast her mother aside. "People suggested that she was crazy to send her children, especially her girls, to school," Hindou remembers. "That was a moment of great challenge for her, but she was resolved to give her children a Western education and identity."

During the school term Hindou lived in N'Djamena, but every summer she would return to the vastness of the Sahel and her community's nomadic way of life. Under her grandmother's direction, Hindou followed the same rites of passage as the other nomadic girls in the community, milking the cows and performing chores. But on her return to school in the late summer, she would become the target of school bullies, who mocked her for her roots. "The other girls did not want to sit next to me because I was from a pastoralist community. They taunted me, saying that I smelt of sour milk. I felt discriminated against." Tired of the abuse, Hindou decided to take matters into her own hands and started an organisation at the school for bullied kids. She was just twelve years old. "It was an association to protect the rights of these other marginalised kids," Hindou said, "and my own."

Two decades later, Hindou now applies that same crusading spirit of justice in speaking out on behalf of the M'bororo community and on rights for women and indigenous groups. She became a climate activist in 2000 while attending a meeting in Nairobi where she heard, for the first time, the words *climate change*. For years, the M'bororo had been struggling to adapt to the changing weather patterns, and their challenges suddenly made sense to Hindou. "I already understood the extremes in our weather patterns but thought that we were the only people battling this alone using our indigenous traditional knowledge. Now I understood that there were other people around the world working on the same issue."

Trying to navigate the corridors of international climate diplomacy proved challenging. "To be an African woman is to stand on the sidelines. To be an indigenous woman is a double marginalisation," she said. In 2006, at her first UN climate change conference in Kenya, when she was just twenty-one years old, Hindou was granted observer status, allowing her only to circumnavigate the negotiations on the fringe. "I was confused," she remembers. "The M'bororo were the ones living the reality, and those seated at the main table were living in big cities. People living in cities cannot know the reality of climate change, of what we are experiencing, I thought. They cannot decide what is best for us."

Like the pastoralist child once shunned in her elementary school classroom, Hindou decided that she could not stand by and watch her community be ignored. Back in Chad she approached her own government to highlight the challenges that the M'bororo faced and started her own organisation, the Association des Femmes Peules Autochtones du Tchad (AFPAT) to work for the rights and environmental protection of the indigenous Peule and to help them to better manage their own natural resources. "In our pastoralist community, you cannot talk about human rights without talking about environmental rights, because we depend on what the environment gives us," she said. Gathering the M'bororo men, women, and elders, she created a three-dimensional map of the community's resources that tapped into their traditional knowledge. The men pointed out where the mountains, rivers, and sacred places were, while the women showed Hindou where they travelled to collect food and water. Afterwards, Hindou invited policymakers from the government to the community to view the map. "We put our 3-D map and the government's satellite map side by side. It clearly showed that our community map had more realistic data on the effects of climate change than the satellite map. We demonstrated that people don't need to go to school to show the government their own environment."

Following the disappointment of Copenhagen, when developing countries accused larger nations of taking the negotiations into their own hands and keeping them shut out in the cold, Hindou persuaded the eminent French documentary maker

Nicolas Hulot to film the plight of the M'bororo in Chad. That documentary, *Espoir de vie*, the last of Hulot's film career, aired in 2011. "That's the moment when my government in Chad began to take an interest in us," Hindou said. In 2013, Hindou was elected co-chair of the International Indigenous Peoples' Forum on Climate Change. She now has a seat at the climate change negotiating table, a significant step up from her wearing an observer badge as before. From this table, Hindou fights to ensure that traditional and indigenous knowledge be a part of any negotiated climate solution. "Indigenous peoples and local communities are on the front lines of this crisis. We bear the brunt of the industries that pollute, and the lands that are left to us bear the brunt of the changing climate. We need to manage our own way forward." Developed nations "have to stop the pollution, to stop the coal mining, and think instead about renewable energy and sustainable development. If they don't think about that, then they are killing us for sure."

Hindou, a striking young woman wearing colourful and elegant African dress, is easy to spot at a climate conference. At the 2016 United Nations climate summit in Marrakech, Hindou was excited because she, Jannie Staffansson, and other indigenous voices had secured a significant breakthrough for indigenous people in the Paris Agreement. This breakthrough, paragraph 135 of the agreement, recognised the need for a Platform for Indigenous Peoples that would be a basis for future action. "We must make progress here in Marrakech with paragraph 135," Hindou said. "It will be a vital platform so that we can get our priorities accepted."

Before the meeting of the Indigenous Peoples' Caucus, Hindou confessed that the group was struggling to make progress with the platform. Invited to address the group, I told them of a remarkable moment in Irish history when the Choctaw tribe in the United States helped the people of Ireland in 1847, the third year of the great potato famine, when the potato crop had failed again. That year, in spring, the Choctaw people met to mark the tenth year of their banishment to Oklahoma from their tribal lands. Somehow aware that millions of people on an island far away were starving, the Choctaw raised $173 at this meeting—a large sum of money at that time—and sent it for the relief of Irish famine victims. One hundred and fifty years later, in March 1997, I travelled to Oklahoma as president of Ireland to thank the Choctaw people. The compassion of the Choctaw for a dying and diminished people thousands of miles away was proof that help can come from unexpected places, and that geography does not have to be a barrier to empathy.

The courage of Hindou's parents in going against the wishes of their pastoralist community to give their daughter an education has inadvertently given the M'bororo community a lifeline in this quiet-spoken but determined young woman. Education has allowed Hindou to become a voice on behalf of her community, to rise above the patriarchal ranks of the M'bororo. It has taken years for Hindou to be taken seriously by the elders in her community, for them to look past her gender, but she has now earned their respect. Recently, while she was taking a group of Western journalists to visit the M'bororo, one journalist asked the elders what they thought of Hindou's work. "She is our hope," an elder replied. "If we see a plane in the sky, we point to it and say, 'Hindou is there at the negotiations. She has gone to get us a solution.'" Such is their belief in her that Hindou's first priority when returning from her climate change work abroad is a trip to the Sahel to visit with the M'bororo elders. She has come to dread the first question they pose: Has she come back with a solution to their problems?

"I tell them that I will have a solution soon," Hindou said with tears in her eyes. "They think that I am finding a solution, but I know how slowly the fight against climate change is going and that a solution is not coming tomorrow. The solution for this problem will not be for them. It will not be for now." Then she steadies herself and shows her strong character. Finding that solution is her goal, and she is determined to fight on, repeating, "We need to manage our own way forward."

The experiences of Hindou and Jannie may seem remote and disconnected from our lives. But their stories should serve as a dire warning. The water shortages facing Hindou's community in the Sahel, and the melting tundra across Jannie's Sápmi, are unmistakable signals of a planet in distress. Their fate is inextricably linked to ours. Their indigenous communities have an intimate relationship with the land and natural world. Years before scientists fully grasped the scale of climate change, herders in the Sahel and in the Sápmi spoke of alarming shifts in the weather. We can learn from their wisdom as they adapt to these tectonic changes. We must listen to them.

Vu Thi Hien left academia to apply her skills on the ground, helping to preserve Vietnam's biodiversity, its ravaged natural forests, and the indigenous communities that live nearby. (Courtesy of Vu Thi Hien)

6

SMALL STEPS TOWARDS EQUALITY

W ITH MORE THAN two thousand miles of coastline, Vietnam is highly vulnerable to the impact of
climate change. The country's low-lying coastal regions and
extensive river delta system make it susceptible to sea level rise
and saltwater intrusion into arable land, particularly in the
Mekong River delta region, where nearly a quarter of the country's population lives. Inland, communities in the upland regions
suffer the effects of more severe and unpredictable weather,
including flash floods, which exacerbate conditions in mountainous villages already rife with impoverishment and food
insecurity.

Vu Thi Hien, a grandmother of four, left a prominent teaching
post at the Hanoi University of Agriculture to help preserve the
country's natural forests and biodiversity, and to support the poor
communities—composed mostly of ethnic minorities—who live
up on the limestone slopes abutting these forests. Ravaged by
overexploitation, Vietnam's natural forests—lush with native
hardwoods such as the *Hopea odorata* and humming to the chorus
of indigenous animals, birds, and insects—have been steadily
disappearing since the 1940s. During the Vietnam War, nearly

eight thousand square miles of forest—6 percent of Vietnam's land area—were obliterated when U.S. forces sprayed millions of gallons of defoliant in a bid to expose the Vietcong hidden among the tropical scrub.[1] Since then, more forestland in highland plateaus has been cleared to make way for higher-value commercial and perennial crops such as coffee, cashews, and rubber, and for shrimp farms and aquaculture.[2] A landmark 1998 decree by the Vietnamese government to restore the country's forest cover to 1940s levels through the planting of millions of hectares (a hectare is nearly two and a half acres) of new trees largely proved successful. But although total forest cover across Vietnam has increased since 1998, an alarming degree of degradation of natural forest has occurred through illegal logging for timber export and paper production, the encroachment of agriculture, and the increasing effects of climate change.[3] Across Vietnam now, only about eighty thousand hectares of primary natural forest remains.[4]

Just like the oceans, which cover a high percentage of our planet, the forests around the world act as "carbon sinks." Throughout their lifetimes, trees take up carbon dioxide and release oxygen through photosynthesis, transferring the carbon to their trunks, limbs, roots, and leaves as they grow. When trees die or are felled, burned, or decompose, the stored carbon is released back into the atmosphere, in what is known as respiration, adding to atmospheric CO_2 levels and contributing to the rise in greenhouse gases. In the fight against climate change, keeping mature forests in the ground is critical not only because they reduce the amount of carbon in the air, but because they are massive reservoirs of stored carbon. Globally, plant and tree respiration contributes six times as much carbon dioxide to the atmosphere as fossil fuel emissions. The preservation of Vietnam's natural woodlands and jungles is critical to the country's fight to lower emissions, as carbon stocks in natural forests are estimated to be up to five times higher than those in planted forests.[5] In the last forty years alone, more than a billion acres of tropical forest around the world—an area equal in size to almost half of the United States—have been razed for timber, mining, development, and subsistence farming.[6] This pace of deforestation is so vast that it now constitutes the second-leading cause of global warming—causing an estimated 15 percent of global greenhouse gas emissions, more than the total emitted by fossil-fuelled cars and trucks around the world. According to Lord Nicholas Stern, Britain's leading climate change economist and the author of the influential review on the economics of climate change, reducing deforestation is the "single largest opportunity for cost-effective and immediate reductions of carbon emissions."[7]

But protecting the world's forests—and reducing carbon emissions—can be successful only if we engage the indigenous communities who live beneath the canopies of our world's forests and who act as their guardians. A 2016 report by a group of academic institutions and environmental NGOs highlighted that indigenous peoples manage at least 24 percent of the total carbon stored aboveground in the world's tropical forests,[8] an amount greater than 250 times the amount of carbon dioxide emitted by global air travel in 2015.[9] At least one tenth of the carbon found in the world's tropical forests is in forests lacking legal recognition, making these forests more at risk from illegal logging or cultivation.[10] As the world begins to turn climate commitments into action following the signing of the Paris

Agreement in December 2015, empowering forest communities to protect their habitats can help to dramatically stabilise rising emission levels. Given that many forest dwellers also rank among the world's poorest inhabitants, helping them manage their forest resources can lift millions out of poverty. Vietnam is home to at least 25 million forest-dependent people, who receive an average of 20 percent of their income from forest resources.[11] The majority of these forest-dependent groups are members of ethnic minorities who live in poverty in the northern uplands and central highlands. But persuading impoverished communities to eradicate logging—when tropical forests are worth more dead than alive—remains a challenge.

In the summer of 1998, after completing her master's degree at the University of Sydney, Vu Thi Hien packed her bags and returned to Vietnam. She had been set on a life in academia, but Australia had stirred something deep within Hien, and in 2000 she abandoned plans for further studies and determined on a new course of action: to apply the skills and methodology acquired in Australia to help people less fortunate than she was back home. "What I learned in Australia was not just academic. I learned how to frame my thoughts in a country with an effective rule of law. By living in Australia, I began to understand what a country needs for development, and how I could contribute to improving life back in Vietnam." Hien had worked as a development consultant on the design of a microcredit project in the upland areas of Vietnam's Phu Tho and Lao Cai Provinces and had been moved by the poverty she had witnessed among the provinces' ethnic minorities, particularly women. Now, Hien instinctively set her sights on the serrated mountainous slopes of northern Vietnam, an area of dazzling beauty but unrelenting poverty. Approximately three hours north of Hanoi, in the Vo Nhai district, Hien found narrow clay roads that led to tiny hamlets hemmed in by sheer limestone mountain walls. In the village of Binh Son, north of the Cuc Duong Commune, Hien encountered members of the ethnic Tay, Dzao, and Hmong groups, their uniquely preserved tribal culture appearing centuries away from the rapid globalisation and urban development sweeping across Vietnam's lowlands. On the edge of the vast tropical forests, the villagers showed Hien how they tended small subsistence plots that yielded wet rice, maize, cassava, and vegetables and raised pigs, buffaloes, and chickens in the shade of the native hardwoods. Each family supplemented its meagre farming income by making daily forages under the tropical-forest canopy to collect arrowroot, yam, brown tuber, and firewood. Village women scoured the damp ground of the forest for precious medicinal herbs such as amomum, *yen mat*, and *linh chi* mushrooms. But in recent decades, the villagers told Hien, the forest perimeters had begun to shrink owing to excessive logging. The majestic and prized *Burretiodendron hsienmu* tree, so vast that several adults linking arms could not reach around its trunk, was now a rare sight. The cacophony of the forest's rich biodiversity—from jungle cocks, white storks, leopards, monkeys, and chamois—had dimmed with the roar of the loggers' chain saws. As each new hectare of forest was cleared, the effects on the surrounding villages became more apparent. "The forest regulates the temperature of the surrounding villages," said Hien. "In the summer, without the shade of the trees, it was too hot; in the winter, too cold. Because the trees were being cut, there was no water flowing down to the farms. Drought became a regular occurrence. The crops in the communities began to fail, and income from agriculture was lost." With so many women in the villages reliant on the forests as a source of income, the degradation of these resources began to drive and exacerbate gender inequalities.

Motivated by her research and the plight of these mountain-based ethnic minority communities, Hien decided to establish her own NGO, the Centre of Research and Development in Upland Areas (CERDA). Through CERDA, Hien had a powerful vision: to persuade the Vietnamese authorities to grant the upland communities formal ownership or usage rights to the forests so that they could act as custodians, generate income, and mitigate the effects of logging and climate change.[12] "I had been moved by the families that I met," Hien said. "The images haunted me. In spite of their extreme poverty, these people were full of dignity. They showed pride when applying for government loans. I wanted to help them."

A year later, in 2005, the United Nations Framework Convention on Climate Change proposed the establishment of an initiative known as REDD (for Reducing Emissions from Deforestation and forest Degradation and the role of conservation, sustainable management of forests, and enhancement of carbon stocks in developing countries) to provide financial incentives for developing countries to stop illegal logging of forests and for countries to offset their carbon emissions by investing in projects that promote sustainable forest management. Through the program, later expanded to become REDD+, local indigenous communities living within or near forests were contracted to manage and protect the forests and to monitor how much carbon was being saved. These indigenous communities had sustainably managed their forests since time immemorial, developing traditional knowledge and practices that had allowed them to adapt to the changing climate. Marshalling their expertise would be vital to the preservation of these forests. In 2009, Vietnam became one of the first pilot countries in the world chosen for the UN-REDD program, and CERDA, under the leadership of Hien and with the support of international funders, initiated a trial project in Thai Nguyen Province to help the ethnic minorities there protect and preserve their natural forests.

Confronting Vietnam's strongly hierarchical decision-making system, which places significant control at the provincial and district levels, Hien knew that she had first to engage with the local authorities and village leaders to win the trust of the community. The village leaders helped Hien hold communal meetings to educate villagers on issues related to climate change and the importance of forest preservation. At these meetings, Hien told the villagers that the knowledge and traditions that they had accumulated over generations to protect the forests could now be used to create income for their families. Persuaded, the villagers agreed to form self-governing groups—each unit its own legal entity comprising fifteen to twenty-five households. Each community group elected a chairman, a board of directors, and a control and management board that included the village heads. Membership in the groups was voluntary, and any villager wanting to join was invited to submit an application for membership. Next, Hien approached the district authority, the agency with the power to allocate ownership of the forest. The district authority agreed to grant these local groups authority to oversee the use of the forest, and to measure the timber volume. From 2012 to 2015, seven cooperatives in Thai Nguyen Province comprising 2,459 households had responsibility for the governance of 4,300 hectares—an area equivalent to three quarters of the size of Manhattan.

With funding from the United Nation's REDD+ initiative to help establish the project, Hien helped the villagers purchase GPS and carbon-measuring devices. CERDA then trained the villagers on how to read maps, use the GPS, create a forest inventory, and measure carbon. Next, the cooperatives developed a plan to protect the forests and began a transparent, inclusive process of allocating land to cooperative members. For a nominal fee, charged per hectare and paid to the cooperative, participating households secured access to an area of the forest. Household members were then organised into teams and sent on regular forays into the forest to either assess the state of the trees or to monitor and report illegal activities such as logging. Through a revolving fund for microcredit tied to the amount of timber and carbon saved, the villagers then received performance-based payments from REDD+. By August 2016, the cooperatives were reporting significant improvements at the borders of the forests and a reduction in the illegal harvesting of trees for export and firewood. "The villagers now say that the water has returned, and in the forest many new plants and animals— including rare monkeys—are springing up," Hien said. "Animals are very wise. They have returned to the forests because they know that the forests are safe. That makes me feel very happy and warm inside."

Vietnam is undergoing a great economic transition, and that rate of change is evident on the road out of Hanoi towards the jagged mountains and cascading rice fields in the north of the country. Along the newly constructed motorway, old farming lots compete for space with large multinational manufacturing plants, and bicycles overloaded with local produce are dwarfed by trucks hauling minerals and goods for export. In the fifteen years following the turn of the millennium, the country's GDP has increased by 575 percent. As the country shifts rapidly from a traditional, agrarian economy to one underpinned by the manufacture of modern goods, the number of people living in poverty has rapidly decreased—people are living longer, and large portions of the population are enjoying a higher quality of life than ever before. But despite the rapid economic growth, local communities still face challenges of poverty and social exclusion.

Arriving in Phu Thuong commune in the district of Vo Nhai, Hien, a slim, elegantly dressed woman whose broad smile lights up her face, introduced me to the district manager, cooperative leaders, and members, many of whom were women. Hien's approach is impressive: Everyone in the room was invited to have his or her say. The district manager of Vo Nhai expressed his pleasure that this area of forest was so well managed and praised the cooperatives for defending their areas with such vigilance. Reflecting on how far the CERDA project in the Phu Thuong commune had come since inception, Hien admitted that she was pleasantly surprised by the progress. By September 2016, almost five thousand hectares of natural forest across seven communes in northern and central Vietnam had been protected. "It is beyond my expectations," she said. "The forest is now jointly managed by the community, and the practice of harmful forest use has been minimised. The volume of wood has increased in the forest as the resource recovers, and the communities have a new source of income from their forest protection payment."

Hien's work to empower self-governing groups to manage the forests has brought about a range of astonishing social outcomes for this predominantly patriarchal rural society. This was most evident in the number of smiling female faces seated in the community hall, particularly the number of indigenous Dao women from higher up the mountain. Prior to CERDA's involvement, women in ethnic minority groups in the upland areas had been shut out of forest management and were denied access to land ownership. But the empowering of women to participate in the forest CERDA programs had fundamentally shifted how women were viewed in the forest communities. "Once the level was right for women to feel they could participate as equals, I knew it would be right for the community," Hien explained. Several women in the room told me how confident and self-assured they felt when engaging with others in their village, and how they now had the courage to participate in village decision-making. One woman from the Ba Nhat cooperative spoke of her sense of ownership and pride, and how the forest cooperative had transformed her life: "At first I was too shy to speak when I joined, but then I realised we could all take part and say what we thought should be done."

"Do you think your daughter will be shy at first like you?" I asked her through translation.

The woman's answer was immediate and strong: "There is no way my daughter will be shy. She has seen me speak out and she knows women have an equal voice."

By putting people and local communities at the heart of forest management, Hien has empowered people living on the very front lines of climate change. "When you work with vulnerable and poor people, you must believe them," she insists.

"Poverty does not equate with stupidity. These people have their own knowledge, their own technology, their own systems. Not only do they contribute solutions, but they are the beneficiaries too. These are the people who can protect and save our planet."

Anote Tong returned to his Pacific island nation following the
2009 Copenhagen climate change conference to tell his
people that Kiribati was in peril of being engulfed by the sea.
(© Matthieu Rytz/anotesark.com)

7

MIGRATING WITH DIGNITY

A S PRESIDENT OF Ireland, I had the privilege of meeting with hundreds of thousands of Irish at home and abroad, and visiting thousands of community groups and organisations across the country. But I never had to return home from an international conference and tell the people of Ireland that our land might soon become uninhabitable because of the onslaught of climate change.

That's what happened to Anote Tong, former president of the Republic of Kiribati, when he returned to his Pacific island nation following the Copenhagen climate change conference in December 2009. He had to tell his people that Kiribati was in peril of being engulfed by the sea. Kiribati—pronounced *keer-i-bas* in the local language—is made up of thirty-three coral atolls and reef islands, located on the equator about halfway between Australia and Hawaii. Scattered across an ocean area the size of Alaska, Kiribati's many islands, home to slightly more than one hundred thousand people, reach at most barely six and a half feet above sea level. The latest climate models predict that melting polar ice and thermal expansion of warming seawater may cause the world's oceans to rise by two to four feet by 2100.

Nearly twenty years ago, because of its position on the international date line, Kiribati was the first country in the world to welcome in the new millennium. Now, in a tragic twist of fate, it may become the first one lost to the effects of climate change before the dawn of the next century.

In response to this threat, in 2014, Tong purchased about six thousand acres of forested land on Fiji's second-largest island, Vanua Levu, one thousand miles away. Five years earlier, the Pacific nation of the Maldives—also threatened by rising sea levels—became the first country to consider moving its sovereign state when the Maldives government looked to India and Sri Lanka for potential land. Tong's decision to spend $8 million on the Fiji land came in the wake of the fifth assessment report of the United Nation's Intergovernmental Panel on Climate Change (IPCC), which confirmed—in the starkest tones yet—that small islands in the Pacific and Indian oceans risked total annihilation. "Coastal systems and low-lying areas will increasingly experience adverse impacts such as submergence, coastal flooding, and coastal erosion due to relative sea level rise," the report stated. On South Tarawa, Kiribati's six-square-mile capital island, where approximately fifty thousand people live, the land is so narrow that it is possible to stand in the middle of the island and see ocean on one side, and the lagoon on the other. In the run-up to the report's release in March 2014, Kiribati had been inundated by a series of extreme "king tides," which sent polluted seawater crashing into people's homes, breaking apart their flimsy dwellings, and sending residents fleeing to higher ground. "Buying land provides a moral sense of comfort that we do have an option," Tong reflected on his decision to buy land in another country as a backup plan for his people. It was also a powerful rebuke to the international community, which was paying little heed, Tong believed, to his nation's doomsday scenario. "The message was loud and clear: Whether you believe it or not, whether you are going to do anything about it or not, our fate is sealed," said Tong. "At some point within this century, the water will be higher than the highest point in our lands."

∞

In the Paris Agreement, countries agreed to "holding the increase in the global average temperature to well below 2°C above levels prior to the Industrial Revolution and to pursue efforts to limit the temperature increase to 1.5°C above pre-industrial levels" as necessary for the world to avoid the worst effects of climate change. But rising greenhouse gas emissions in recent decades have made achieving this target difficult, making it possible that the global temperature will rise to 3°C or 4°C above pre-industrial levels within this century—a catastrophic outcome. In the spring of 2015, scientists discovered that quantities of carbon dioxide, methane, and nitrous oxide from industrial, agricultural, and domestic activities had reached record levels, crossing the four-hundred-parts-per-million guardrail for the first time. That same year, a strong El Niño—a periodic, natural variation in sea-surface temperatures—developed in the Pacific Ocean, impacting global temperatures. By the end of 2016—the hottest year on record to date—the global temperature increase had already surpassed 1°C above pre-industrial levels.

While world leaders struggle to reduce global carbon emissions, many observers see Kiribati as the proverbial canary in the coal mine: a real-time precursor of how rising seas and intensifying storms threaten the existence of an entire nation. During a speech at the United Nations in September 2016, the head of the International Organization for Migration, Bill Lacy Swing, warned that climate change threatens a staggering seventy-five million people around the world who currently live just one meter or less above sea level. In parts of Florida and Georgia in the United States, the accelerating rise of the sea is already causing frequent tidal inundations, where a mere high tide—known as sunny-day flooding—can force seawater over the top of man-made barriers and into the streets below. Millions of people living in developing countries at low altitudes, particularly those along the coast of Asia who have no such barriers to protect them, are likely to lose their homes as flooding and rising sea levels sweep the region. In Bangladesh, scientists predict that by 2050 as many as twenty-five million people could lose their homes and livelihoods to the rising sea. In Africa, where more than 25 percent of the population live within a hundred kilometres of the coast, three hundred million people are at risk from flooding caused by sea level increase.

In 2013, a World Bank report, *Turn Down the Heat*, laid out in shocking detail the ravages that Kiribati's residents—who as their coastal villages become uninhabitable have no options to retreat inland because they have no high ground—should expect in the coming decades. "What is the future for us? The reality is that we won't have a home," Anote Tong said. "The IPCC projection is for the global sea level to rise by about a meter by the end of the century. I know exactly what that means for us. It won't be by the century. It will be well before that." Kiribati's socioeconomic trends—high population growth rates and migration to the capital from outer islands—have exacerbated the atoll's vulnerability, while poverty, overcrowding, and poor sanitation have begun to deplete the island's already-limited water resources. Add climate change into the mix, and with significant sea level rise, Kiribati's freshwater supply will be even more imperilled. With alarming changes in weather patterns, flooding has in recent years become the norm.

As a human rights advocate, I am reminded constantly of Eleanor Roosevelt and the commission that drew up the Universal Declaration of Human Rights, adopted by the General Assembly of the United Nations in 1948. In remarks delivered at the United Nations in March 1958, Roosevelt remarked how human rights begin in small places, close to home, some so small that they cannot be seen on any maps of the world. "Unless these rights have meaning there," she said, "they have little meaning anywhere." A mere seventy years after the adoption of what Roosevelt referred to as the international "Magna Carta," this remarkable woman could never have envisaged that human-induced climate change would cause such devastation to poor communities and threaten the existence of sovereign states such as the Republic of Kiribati. "As a child, I used to visit a remote village island some distance from my home," Tong told me. "But later, during my lifetime, the village began to disappear. Several years ago, the water rushed in and now the village is no longer there." All that remains of the village is an old church that juts out of the Pacific waters like the tip of a modern-day Atlantis. In recent years, Tong asked villagers to build a seawall to protect what remains of the church, a stark visual reminder to the former leader of his race against time.

Tong is a slim man of Chinese and Kiribati heritage, with a trim moustache and grey crew cut. Born on one of the outer-lying islands of Kiribati, south of the tiny capital island of South

Tarawa, Tong escaped the extreme poverty of his childhood when he was sent to New Zealand at the age of six to attend a Catholic boarding school. Educated by Irish nuns, Tong remained in New Zealand for the rest of his childhood, eventually graduating from Auckland University with a degree in chemistry. After some time spent working for the Kiribati foreign service in Fiji, Tong returned to Kiribati in the 1970s. In 2003, following a contentious political fight against his older brother, Harry, and with a margin of just one thousand votes, Tong was elected president.

An accomplished fisherman who knows intimately the contours of many of Kiribati's sandy atolls, Tong noticed decades ago that something was amiss with the weather. He began to research assiduously the science behind climate change, taking into consideration both sides of the debate. "Like a lot of people around the world, the ongoing controversy about the science threw confusion into the argument," he remembers. "We heard what was being predicted but still hoped that there was a chance." But in 2007, everything changed when, four years into his presidency, Tong read IPCC's fourth assessment report. "After that, I really began to panic," Tong told me. "I started to read more detail and began to analyse what it meant, not in terms of the science, but for the people of Kiribati. I became very frightened." Immediately, Tong set about drawing up contingency plans for his nation's demise. But concerned by how the news might be perceived among his people, for the time being he said nothing. "It was a deliberate ploy on my part not to tell my people at this time, because nobody could grasp the enormity of what was at stake. I knew that there was nothing that they could do to stop what was happening. All I would end up doing would be to make the rest of their lives miserable."

∞

In December 2009, the United Nation's annual climate change summit in Copenhagen was highly charged, marred by scenes of chaos and recrimination in the closing stages. Although Copenhagen produced the first joint commitments on emissions by major developed economies, these cuts fell far short of what Tong and leaders of other vulnerable countries had been holding out for in hopes of keeping the global temperature rise to 1.5°C in this century. In the dramatic last minutes, all references to 1.5°C in the draft deal—brokered between China, South Africa, India, Brazil, and the United States—were removed, blindsiding Tong and those from other small and more vulnerable nations. The message was clear: In the global fight to reduce greenhouse emissions, Kiribati and other Pacific nations would become collateral damage. It was a slap in the face to Tong. His presidential colleagues on the world stage—the leaders of developed countries that had built their economies on fossil fuels—had signed Kiribati's death warrant.

It was painful to witness a group containing the leaders of a number of smaller nations literally shut out from the negotiations and standing in the December snow while the major world powers argued late into the night. I began to understand that the fight for climate justice had to be centred not just around the individual, but at the state level too—that the concept of climate justice must be broadened to ensure that smaller states are given a voice and a place at the negotiating table.

For Tong, the UN climate summit meeting at Copenhagen was a watershed moment. He returned to Kiribati alternately dejected and furious, knowing that if he was to save his country, he needed to act alone. "I was very angry," he remembers. "I now understand that people can become extremists when they are not being heard. After Copenhagen, I had a deep sense of betrayal and a sense of futility. But I knew that I had to keep thinking that there was something more that we could do. If not, then I did not deserve leadership." For weeks, Tong could not shake an image in his head, that of the Kiribati people floundering in the water and struggling to board a life raft captained by developed nations. At Copenhagen, Tong had felt patronised by the leaders of some developed nations who had lectured him on the danger to their economies of an agreement containing anything less than 2°C. The rebuke stung. "I told them that anything above one and a half degrees centigrade was dangerous to our future as a people. When the time comes, I am haunted that the people of Kiribati will be pushed away as they try to scramble onto the life raft."

In the weeks following Copenhagen, while Tong drew up plans for his country's demise, he tried to grapple with a profound sense of failure and helplessness. "Then I had to step forward and say we have to find a solution. I had to acknowledge, to come to terms with, the reality of what was happening." Next, Tong tried to understand how to broach the topic to a sceptical audience not only at home, but also abroad. "I had to change how I was talking about the issue. You don't listen to somebody who accuses you of doing the wrong thing. I had to make other countries realise that it's not just my problem. That it's their problem too, whether they like it or not. But even then, it still took a long time to be heard."

Anote Tong's dignified manner, his quiet authority, his moral bearing, make clear he was no ordinary president. "The science on climate change is very clear. The severity and urgency of the threats may not be the same for each of the countries represented in Copenhagen that day, but the direction is unquestionably the same." Going forward, Tong promised, he would do all that he could on the international level to save his country, but he would also rely on the strategy that had served him best until now: to tell his story. Just like Constance Okollet, Tong instinctively recognised the power in truth telling and was determined to take Kiribati's plight on the road. "I will tell my story to anyone who will listen. I want to emphasize the human dimension, which has been ignored until now. I want to talk about how there is attention on polar bears, but none on the people in the middle of the planet."

In the wake of Copenhagen, Tong's story would catapult Kiribati into the international media spotlight. But the attention would bring him criticism back home from local opposition and Christian leaders, and some residents, who saw their fate as resting only in God's hands. Some international scientists also disagreed with Tong's pronouncements about erosion and flooding, suggesting that he was prone to overstating the role that sea level rise has played so far. But each year the pull of the moon brings new king tides to Kiribati, and the seawater contaminates drinking water, ruins crops, and causes food shortages. Tong recognises that the more he talks about climate change and what the future holds for the young people of

Kiribati, the more that people will want to leave the islands. To enable the future of these young people, he has created training programs in a variety of skills, including nursing and carpentry, to give the youth of Kiribati an economic lifeline for when the islands become uninhabitable. "Migration with dignity is a real strategy," he said. "I think what is new about this concept is its application to climate-induced migration. I want migration from our country to be a painless process, even a happy process, for those who choose to go. They will go on merit. We will prepare them."

In 2016, after three terms as president and no longer eligible for reelection, Tong stepped down from office. He was anxious about relinquishing his post as president, knowing, he says, that in spite of the new Kiribati government's support for measures to combat climate change, it was not at the top of their priority list. "That is why I must keep talking with them. We can never give up. We can't afford to give up. No matter what the obstacles seem to be. For us there is no giving up."

Tong has now engaged with Japanese, Korean, and United Arab Emirates engineering firms who are leading the race to create giant floating islands. He has yet to work out the daunting financial and logistical details, but hopes that artificial floating islands may offer a lifeline for Kiribati, if not for this generation then for the next.

When people are drowning, Tong believes, they will grasp at anything to stay afloat. That is why he will devote the rest of his life to the Sisyphean task of saving the Kiribati nation.

"I'd rather plan for the worst, and hope for the best," he observes.

Much of Natalie Isaacs's working life revolved around generating plastic waste. "I used to think, 'Well, what can one individual do anyway? It's not my problem.' I felt that way for a very long time." (Photo by Jenny Kahn)

8

TAKING RESPONSIBILITY

As a successful businesswoman, Natalie Isaacs thought that she knew a lot about climate change, at least enough to hold her own during dinner-party conversations in the well-informed social circles that she and her husband navigated in Sydney, Australia. But with her own skin-care and beauty products company to manage, and the busy lives of her four children to run, little space was left in Natalie's crowded life to get involved in environmental issues, let alone enough time to recycle the family's paper and plastic waste.

"I could sit around the dinner table and talk about climate change and express horror at what was going on and then just leave and carry on as normal," Natalie—with abundant curly red hair and a warm, cheerful voice—remembers. "I didn't even have recycling sorted. There was a real disconnect between being aware about what was going on, but not actually doing anything about it. I used to think, 'Well, what can one individual do anyway? It's not my problem.' I felt that way for a very long time. Climate change just wasn't my issue."

Competing in the aggressive skin-care industry, much of Natalie's working life was spent generating plastic waste. At business meetings with her staff, Natalie schemed and plotted about how best to outmatch her closest competitors, to win over buyers with attractively parcelled beauty products, plastic tubes, and containers swathed in cellophane packaging. "My entire working career was about how I could get someone's product off the shelf, so I could replace it with mine," she said.

But in 2006 a series of random events forced Natalie to pause and contemplate the changing world around her. That year, bushfires—some of the worst that Australia had seen in decades— descended upon Sydney, the flames licking houses at the edge of the city and blocking out the sun with acrid black smoke. Major roads leading north out of Sydney were shut as thousands of firefighters battled to contain the fires. Sydney had experi- enced similar fires four years earlier, but this time it all felt a little too close for Natalie, who watched the encroaching flames with increasing alarm from her home on the northern beaches of the city. "The drought, the heat, and the fires seemed the worst they had ever been that year," Natalie recalled. "That seems like nothing compared to the extreme weather patterns and drought that we are dealing with now."

That same summer, Natalie watched Al Gore's landmark documentary, *An Inconvenient Truth*. She listened as Gore outlined the "moral imperative" of addressing climate change and watched with horror as he displayed graphs crisscrossed with alarming red lines showing increasing rates of carbon dioxide emissions and the corresponding rise in temperatures. None of this was new to Natalie's husband, Murray, a former environmental reporter who was now working as a sustainability consultant and writing a book on climate change. Submerged coastlines, receding ice fields, and changing weather patterns constituted

Murray's daily bread and butter. "Murray would occasionally ask me to read and edit his book," she said. "There was a lot of information in there that I did not understand. I would read a passage and ask Murray, 'What does three degrees of warming actually mean?' Suddenly, with mounting horror, I began to understand."

That summer of 2006, Murray worked with an environmental organisation that engaged university student volunteers to install free energy-efficient lightbulbs in people's homes. The start-up needed some marketing and selling expertise, and they approached Natalie for her help. At a party one night in Sydney to celebrate the placing of the millionth lightbulb, an idea ignited in Natalie's mind: "All of the various messages that I had been receiving about climate change began to add up. Right there and then, I decided that I was going to get my electricity bill down."

The next day, Natalie decided to replace all of the lightbulbs in her home with more energy-efficient models, and to switch these lights off whenever she left a room or the house. Next, she unplugged any electrical gadgets that were not in use—her washing machine, the television, and the stereo—and began to hang clothes out on a line to dry in the summer heat. Within a short time, her family's household electricity consumption had decreased by 20 percent. Holding the electricity bill that came in the wake of her energy-saving efforts, Natalie felt a shiver of excitement. "That was the moment I had an epiphany. I changed forever. I took ownership of the issue. I said, 'I am going to change the way that we live.' I wanted to do more. I wanted to do it all."

Natalie then turned her attention to her household waste. She began to buy fewer plastic and paper products—plastic wrap, freezer bags, paper towels—and created a diligent recycling regimen. She bought more locally sourced produce, and less meat. She avoided foods that came with heavy packaging, created a compost heap, and bought a couple of worm farms. Within a couple of months her household waste was down nearly 80 percent. The feeling was addictive, emboldening. "When I did one thing and saw the result, that empowered me to do the next thing, and the next, and the next. I was learning the secret to life: that less is more. It was such freedom. I felt so light."

∞

Extreme weather patterns have become standard in recent decades across Australia as drought, bushfires, floods, and heat waves hamper the economy of a country with one of the highest carbon footprints in the world. Since the 1970s, northern Australia has become wetter, the south has become drier, and violent bushfires have scorched millions of acres across the country. In 2016, Australian government scientists released a report that showed that the surface temperature across Australia had increased by 1°C since 1910.[1] Although this rise may seem small, it is enough, scientists say, to shift baseline averages and to destabilise the weather and cause more extreme weather events. Drawing on data from Tasmania, where greenhouse gases in the atmosphere are measured, the report concluded that rainfall across some regions had dropped by almost 20 percent. The lack of precipitation has wreaked havoc on Australia's agricultural productivity, with many farmers—particularly those in the Murray–Darling River breadbasket basin—struggling to cope with changing weather patterns and decades of persistent drought. In early 2017, heat waves across New South Wales pushed temperatures in excess of 45°C (113°F), causing concern that such record-breaking numbers may become the norm. Three years earlier, in 2014, a heat wave forced organisers to temporarily shut down the Australian Open tennis tournament when the thermometer reached 44°C (111.2°F).

Up to 75 percent of ocean warming has occurred in the southern hemisphere, which is in the crosshairs of climate change. These warming seas have had a devastating impact on one of Australia's national treasures, the Great Barrier Reef, which spans fourteen hundred miles off the northeastern coast of Queensland. The largest living structure on the planet, the Great Barrier Reef consists of twenty-nine hundred smaller reefs and more than nine hundred islands, all of which are slowly bleaching in the tepid ocean waters. Using aerial and underwater surveys, marine scientists in 2017 declared that huge sections of the reef—stretching across hundreds of miles—were dead,[2] killed off by warmer water. Scientists had not expected to see this level of destruction at the reef for another thirty years. Farther south, the koala bear—Australia's iconic marsupial—is once again facing the threat of extinction as temperature increases, drier spells, and extreme bushfires destroy the animal's natural habitat. Environmentalists warn that this cuddly symbol of Australia could suffer huge reductions in numbers unless the government implements changes to its land-management practices to curb the effects of climate change.

With so much devastation occurring on this island continent, one might assume that Australia would be at the cutting edge of mitigating carbon pollution. But the country remains addicted to coal, the burning of which is the single largest source of greenhouse gas emissions. In 2016, Australia was the world's

leading exporter of coal. That same year, the Australian government gave the go-ahead to the Indian-owned conglomerate
Adani to open the single biggest coal mine in the country. The
proposed $12 billion mega-mine could be in operation for sixty
years and create 4.7 billion tons of carbon dioxide over its
lifetime, dramatically increasing global carbon emissions. In
June 2017, a group of prominent climate change experts and
oceanographers wrote to Australian prime minister Malcolm
Turnbull urging him to reject the Adani mine proposal, saying
it would have a devastating impact on the Great Barrier Reef.
In addition, countless other proposals to create new mines are
currently under consideration in the Australian capital, Canberra.
If successful, these proposed new mines will more than double
Australia's coal exports. Australia is now caught in the most
Faustian of bargains: to see the economy thrive in the short term
through the mining and burning of coal, its most abundant
resource; or to preserve unique treasures such as the Great Barrier
Reef by turning its back on fossil fuels.

∞

By 2008, Natalie Isaacs was struggling with the next—and most
ambitious—stage of her carbon-free plan: getting rid of her
petrol-guzzling four-wheel-drive vehicle, a car she had bought
just two months before her lightbulb epiphany. Living in the
northern coastal suburbs thirty kilometres from Sydney, Natalie
relied heavily on her car and never took public transportation.
"Getting to the city was such a hassle by bus," she said. "I always
drove. It took me a couple of years before I could bear to part
with my car. But once I finally got on a bus, it wasn't so hard.

Taking public transport is now the first thing that I think of, not the last."

Two years into her new energy-saving lifestyle, Natalie had begun to think on a larger scale. Reading up on the level of greenhouse gas emissions for the residential and commercial sectors, Natalie discovered that nearly 17 percent of greenhouse gas emissions were the product of 1.5 billion households around the world. These household emissions were caused by fossil fuel combustion for heating and cooking needs, electricity consumption, waste management, and leaks from refrigerators in homes and businesses. In the United States alone, individual households were responsible for 4 percent of global emissions, with the country responsible overall for 14 percent of global emissions. The apparel industry, generator of the garments and accessories that Natalie loved so much, was the second-largest industrial polluter, behind oil, and accounting for 10 percent of global carbon emissions.

As a businesswoman, Natalie understood that women were influential consumers, but could they also be powerful agents of change? Natalie knew nothing about running a movement, but could not resist a thought that refused to go away: Could she use the power of women in individual households to fight climate change? "If I could just bottle up what had happened to me and share that with every woman that I knew, perhaps these women would want to change the way that they lived?" But how many women would it take to make a difference? A thousand? Ten thousand? One million? That's when the idea struck: "I saw that I had saved all this money and helped curb pollution. I thought, 'Wow, if I could do that just by being a little bit more vigilant around the house, imagine if millions of us were doing that? What would happen if we all cut our meat consumption just by fifty percent? Or if we got our electricity down by twenty percent? Or bought fifty percent less 'stuff'? If somebody just does it on their own, you think, what difference will it make? But if whole communities do it—if the entire population lived differently—it changes the system. There is so much power in actions like lifestyle change because not only does it cut pollution, it also helps you to find your voice."

In 2009, Natalie stepped away from her beauty-products career and started 1 Million Women, an online movement to encourage women to cut back on their carbon emissions. Using a cleverly designed website, Natalie provided simple tips to help other women lower their household carbon footprint. Using a simple "dashboard" on the website, 1 Million Women members logged on to record their energy-saving efforts each week— noting how often they had turned off electrical appliances, bought local produce, recycled, or hung their laundry out to dry—and then received a readout of their personal carbon footage. Natalie's message and website were specifically targeted towards women of affluence living busy lives in consumer-driven cities and suburbs. "If women and children of developing countries are the most vulnerable to climate change, then women from wealthy countries have so much to contribute to the solution through the way we live," she said. "This is about lifestyle."

Natalie struggled for some time to gain momentum with the 1 Million Women movement. What was so easy to accomplish within her own home proved difficult to translate to a wider audience. Motivating behaviour change in Australia's high-consumption society was harder than it looked. "I thought if I could just bottle up what had happened to me and share that with every woman that I knew, then everyone would want to change the way that they lived," she said. "But it didn't happen quite like that. I thought we would have a million women in six months, but it took a really long time because behaviour change is the elephant in the room."

As Natalie worked to convey her message, many women complained that they couldn't possibly fit yet one more to-do item into their busy lives. Natalie, who in a short time had progressed from decreasing her electricity bill to starting a global movement, found their criticism hard to take. "I thought everyone should share my enthusiasm," she said. "But I now understand that people view looking after the environment as an adjunct activity. It's always separate from how we live. If you think it is separate, of course you are going to wonder how can you fit it in?"

Natalie began to address, one by one, the lifestyle changes that 1 Million Women members were finding hardest to tackle, and to tailor her website around these needs. "That is what 1 Million Women is all about. That we are not perfect. This movement is not meant to make you feel guilty. It is meant to make you feel inspired that we are all doing this together, that we're just all trying to do our best. You can't just go into total despair over the effects of climate change. You've just got to start and do one thing, and that leads to another, and before you know it, you are just living like that."

Like many 1 Million Women members, I struggle to lower my own carbon footprint one item at a time. I am aware that air travel, critical for the success of our climate-justice agenda, has a significant carbon footprint. I use too much paper, both at the office and at home. Unlike my youngest child, Aubrey, I am not a vegan. I am slowly taking steps towards becoming a vegetarian and eat less meat than I used to. I now assess which meetings I can attend by video conference rather than in person. My foundation mitigates my own carbon footprint by measuring the amount of greenhouse gas emissions that I expend through air travel and offsetting it through annual charitable donations to a climate-change-related organisation.

Like Natalie Isaacs, and the hundreds of thousands of 1 Million Women members, I learn as I go. But by undertaking this journey to reduce our carbon footprint, we can participate in a global movement that has a real capacity for change. When faced with the enormity of the climate change problem, it is easy to throw our hands up in the air and admit defeat. But individual empowerment leads to confidence. "It is so much easier to do nothing, but you've got to get over that when it comes to climate change," said Natalie. "Just get on with it and do something. At 1 Million Women we work on giving women bite-sized chunks with very tangible results. It doesn't matter what it is. Do one thing, see a result, and that will lead to something else."

As of mid-2017, Natalie's movement has grown to more than six hundred thousand around the world, including a small percentage of men, and is steadily growing. With many members now accessing the 1 Million Women website on their smart-phones, Natalie is creating a new mobile phone application that will track not just how effectively each member is reducing her own carbon footprint, but also the entire global member-ship. With one swipe, members will be able to see how much pollution the 1 Million Women community has saved on any one day, in any location around the world. Plans are in place to create a currency for the carbon pollution that 1 Million Women save—a goodwill carbon—that can be given back to women in developing countries. "We want to emphasise that our individual daily actions are supporting our sisters on the planet who are feeling the effects of climate change right now," Natalie said.

When Natalie first started out on her climate-justice journey, she maintained her inspiration by keeping the fate of her four children, and her one grandchild, uppermost in her mind. But as bushfires continue to rage across Australia, and her government maintains its dependence on coal, Natalie has revised her long-term vision. "This is no longer just about my children's children. This is about us. It is all happening now. Countries are sinking and yet no one country is taking this issue seriously enough." That is why Natalie continues to build her global movement to empower women everywhere to act on climate change through the way they live—to influence through every dollar they spend, and every lifestyle choice they make. How we live each day matters, and one small action at a time multiplied by millions can change the system.

"I am the eternal optimist. I know we don't have much time in the race against climate change, but humanity can rise to this," Natalie said. "If a million of us did this one small thing, it could change the world. A change in lifestyle doesn't need any policy. We just have to keep on fighting. It literally just takes one person at a time."

With the closure of the Brunswick Mine looming, union leader
Ken Smith and 1,500 miners faced a precarious future.
"I realised that a big piece of my identity had been taken away."
(© Mychaylo Prystupa)

As head of the International Trade Union Confederation (ITUC),
which represents 181 million workers worldwide, Sharan Burrow
leads the global movement for a Just Transition towards
a decarbonised world. (© Horst Wagner/ITUC)

9

LEAVING NO ONE BEHIND

I N T H E W A N I N G days of April 2013, groups of men clad in navy overalls and hard hats stepped into the boxy elevator cages of a mine in the province of New Brunswick in eastern Canada and descended into the belly of the earth for the last time. Among them was Ken Smith, who had spent the last thirty-three years of his life at Brunswick Mine, working a range of jobs from diamond driller to hoist man to heavy-equipment mechanic. Born in 1961, with salt-and-pepper hair and a warm smile, Ken was just nineteen when he first began work at the underground lead–zinc–copper mine. For nearly fifty years, Brunswick had been one of the world's largest and most profitable ore mines, employing more than seven thousand local people and sustaining the Bathurst community.[1] Now, as the countdown to the closure began, Smith and fifteen hundred miners contemplated the loss of their jobs, and the precarious future that yawned in front of them.

As a union leader privy to management discussions, Ken Smith had known for many years that the Brunswick Mine was likely to close. A global market slump for zinc ore had caused Xstrata Zinc to announce in 2006 that the mine would likely be shuttered at the end of the decade. As a union representative, Ken knew that he needed to use every means he could to stop the closure or delay it. His union immediately negotiated an extension with Xstrata management, a pact revisited in 2009 that secured a five-year lifeline for the miners and their families. But as Ken watched a sister mine nearby close, he knew instinctively that the writing was on the wall for Brunswick. While trying to keep hope alive, Ken turned his attention to a more depressing reality: preparing for closure while also developing a plan B "transition" to help miners who had spent their working lives beneath the crust of the earth find work in new professions. "By 2009, we were working full bore on transitioning our workers while at the same time trying to keep the place open," Ken said. "It was a two-pronged approach: Let's prepare for the worst, and if the worst doesn't happen, then that's great. But if it does, and the mine closes, at least we will be prepared."

To begin, Ken and his colleagues negotiated closure language in their agreement that included hundreds of thousands of government and company dollars to be used to develop a transition plan and to run a transition centre to provide support for workers who might be laid off, to help train miners in new skills, and to source new employment to the Bathurst area. With this money, Ken organised job fairs with nearly a hundred companies hungry to acquire the skills of the Brunswick miners. For miners who had spent their entire lives working at Brunswick, or who lacked second-level education, the union provided professional résumé support and training. With no trade certification for mining available in Canada, Ken and his union colleagues—with the backing of a powerful international union—lobbied every level of government. Their efforts led to the creation of a national program to recognise and certify the skills and competencies of mining workers. "That was a huge victory," Ken said. "We were able to get our skills recognised and documented, which made us potentially employable at other companies in Canada."

As the countdown to the mine closure approached, Ken and his union colleagues felt satisfied that they had done all they could to cushion miners who would soon lose their jobs. Believing that they had covered every contingency, the union even provided on-site mental health services for those who needed extra emotional support. "We had psychologists who came to the mine to provide counselling to the miners," Ken remembers. "One psychologist told me that I would miss this part of my life once the mine closed, that being a miner was part of who I was." Ken resisted the psychologist's assertion, silently believing that she was overreaching. As Ken saw it, a lot more was wrapped up in his identity than just life at the mine.

But when the last ore deposits were brought to the surface of the Brunswick Mine on April 30, 2013, and the smokestacks of the smelters fell silent for the last time, Ken immediately understood that he had underestimated what the loss of Brunswick would mean to him and to the local community. Despite their best efforts to ensure that no miners would be left behind, the new employment and financial opportunities in Bathurst Ken had hoped to generate never materialised. Instead, unemployment in the town skyrocketed, and the multitude of small businesses that had thrived around the mine collapsed. "Bathurst is a ghost town now," Ken said. "Five years ago, if you drove down the main street in the evening, it would have been bustling and busy. Now there are very few cars on the street. I thought that we had done a great job of bringing in companies and running a transition centre to help people find work. But it turns out that we missed that financial piece of the puzzle."

Days after the closure, Ken Smith and his wife packed up the home they had lived in for more than thirty years and joined the exodus to the west, driving nearly five thousand kilometres to the boomtown of Fort McMurray, the home of northern Alberta's oil sands. Ken was lucky: His new job as a heavy-equipment mechanic would pay twice what he had earned back in Bathurst. But leaving the only town and home that he had ever known turned out to be overwhelming to a man who had built his entire life around the local mine. "That psychologist was right, and I was wrong," Ken said. "When the Brunswick Mine closed, I realised that a big piece of my identity had been taken away."

∞

In many countries when people turn on the television, flip on a light switch, or plug in their cell phone, they are almost certainly using electricity generated by a coal-, gas-, or oil-fired power plant. But as our erratic weather and climate patterns prove, relying on fossil fuels to provide electricity is dangerously unsustainable. Seventy-five percent of all greenhouse gas emissions are from the burning of fossil fuels and the methane released during their extraction. If we are to stabilise our climate—and save our planet—we need to dramatically cut our reliance on fossil fuels and instead massively scale up investments in renewable energy sources—wind, solar, small-scale hydropower, geothermal, and low-emissions bioenergy. These investments will generate millions of new jobs and opportunities around the world. Renewable energy already employs more than three quarters of a million Americans,[2] and jobs in the solar and wind industries are growing twelve times faster than are jobs in the rest of the U.S. economy. Globally, nearly ten million people are employed by renewable energy, over half of whom live in Asia.[3] In early January 2017, the Obama administration's Department of Energy released a report[4] showing that in 2016 the solar industry in the United States employed many more Americans than coal, while wind power provided in excess of one hundred thousand jobs. Data released in 2014 showed that coal employed nearly seventy-seven thousand people in the United States that year, a figure that has since been declining. Across Europe, jobs in renewable energy already outnumber those in coal industries. In the United States, solar power now employs more than oil, coal, and gas combined. A 2017 study by the International Renewable Energy Agency[5] showed that investments in renewable power and energy efficiency would add nearly 1 percent to the global gross domestic product by 2050—that's a boost of $19 trillion, not to mention millions of new jobs. The facts are undeniable: The renewable revolution is already under way and is economically effective and potentially more inclusive.

But as we make the transition to cleaner energy, we must remember the millions of fossil fuel workers around the world who have spent their lives extracting the fuel that has fed our economies. They too are victims of climate change and deserve to be treated with dignity. Their story is part of the struggle for climate justice. Others working in energy-intensive industries—steel, iron, aluminium, power generation, and road transportation—will also be affected by carbon reduction and elimination. Although the move to clean energy will open up new positions and financial opportunities, it will not replace all of the jobs lost when the fossil fuel industry finally winds down. We have already seen this trend in the coal-mining areas of America—in Appalachia and the Powder River Basin of Wyoming and Montana—where changing markets, a glut of inexpensive natural gas, and new climate change regulations have made coal less appealing. Among the many voters who came out for Donald Trump in the U.S. presidential election in November 2016 were workers from the dying coalfields of West Virginia and Wyoming who felt stranded by a perceived "war on coal" by President Obama and the push towards renewable energy. Seduced by a candidate who placed coal at the centre of his campaign, they loudly voiced their economic anxieties at the ballot box, desperate not to be thrown on the scrap heap as their industry faces an increasingly grim future.

Ensuring that no fossil fuel workers or communities get left behind in the shift towards cleaner energy is the goal of the movement known as Just Transition. It is a people-first approach, arguing that workers should be given wage and benefit insurance, income support, and access to health care as they move from the fossil fuel sector into the clean-energy and other sectors. Most critically, it means taking care of communities like Bathurst, New Brunswick, which owe their very existence to the mining industry. Pioneered by the late American labour leader Tony Mazzocchi, the Just Transition movement began in the 1970s at the height of the ban-the-bomb movement when Mazzocchi, then a leader of the Oil, Chemical and Atomic Workers Union, recognised that disarmament would cause massive job losses among atomic workers. A beneficiary of the hugely successful GI Bill, which had helped stabilise the U.S.

economy in the wake of the Second World War, Mazzocchi argued that federal assistance should be provided to atomic workers to ease their transition from the nuclear industry. Decades later, when it became clear that the planet's warming was caused by fossil fuel emissions, Mazzocchi called for a "superfund" for workers displaced by environmental-protection policies, noting, "Paying people to make the transition from one kind of economy—from one kind of job—to another is not welfare." Fossil fuel workers who had helped shore up the world's economy deserved, in Mazzocchi's words, "a helping hand to make a new start in life."

Helping decommissioned fossil fuel workers find new work is now the mission of Sharan Burrow, who as head of the International Trade Union Confederation (ITUC) leads the global movement for a Just Transition towards a decarbonised world. With an admirable lack of airs and graces, Sharan is equally at ease with the chancellor of Germany as she is with a miner from her native Australia. She is also keenly aware that her decisions will affect the lives of tens of millions of ordinary people, among the 181 million workers worldwide represented by the ITUC.

Like the majority of the workers who make up the ITUC, Sharan grew up in a blue-collar family. The first woman in her family to attend university, she began her career as a history and English teacher in a local secondary school in New South Wales. As a fourth-generation member of a proud workers' union family—Sharan's great-great-grandfather had won notoriety in the Australian shearers' strike of the 1890s—Sharan was certain to join her local union chapter. Commuting between her classroom and union meetings, amid the backdrop of the Vietnam War and the antiapartheid struggle, Sharan found it hard to resist the call of activism. "The 1970s were heady times for a young woman," Sharan said. "I didn't realise it back then, but those years taught me that even though the struggles we faced might be local, the framework, the policy structure for human rights and justice, is international. I apply that same principle now in the mission for a Just Transition."

Sharan had planned to stay in teaching for her entire career, but in 1986 she was asked by the New South Wales Teachers Federation to temporarily leave the classroom to help in a relief organising role. Although she would briefly return to her teaching post, that short-term job soon led Sharan to be promoted to senior vice president of her state union. By 1992, she was president of the Australian Education Union and in 2000 became the president of the Australian Council of Trade Unions. As she rose to each new level of organising, Sharan applied the principles of social justice that she had acquired as a teacher back in the classroom. "Teaching," she once said, "taught me humility and exposed the inequalities of our world."

Now Sharan crisscrosses the globe urging governments to allow unions and workers to take the lead in the Just Transition global movement, to integrate their voices as the fossil fuel industry winds down. She has worked to convince businesses of the need for both a just transition and a climate-justice approach to climate action. Sharon and I are honorary members of a group of business leaders called the B Team.[6] In January 2015, at the World Economic Forum in Davos, Switzerland, the B Team members committed to reaching net-zero greenhouse gas emissions in their companies and supply chains by 2050. Now the B Team has been joined by Christiana Figueres and supports her Mission 2020 to turn the tide on the devastating impacts of carbon emissions by 2020.[7]

Sharan never loses her focus. "A just transition for us means that unions and workers—at every level—are involved in the timing, the industrial negotiations, and the planning for the massive industrial transformation that must take place," she says. "There needs to be a plan in the most vulnerable places—where coal mines, coal-fired stations, fossil fuel production, and manufacturing are shutting down." She points to the South Australian town of Port Augusta, where decommissioned-coal-plant workers took the lead in planning their future after the 2016 closure of the coal-fired power stations that had bankrolled the desert community. Knowing that the plants were on borrowed time, for five years prior to the closing the local citizens, council, businesses, coal-plant workers, and their union came together and devised a plan. Despite an early closure, the end proposal to replace the condemned coal plants with a solar thermal plant that will create eighteen hundred jobs and save five million tons of greenhouse gas emissions has been delivered. The plan[8] was based on research showing that a solar thermal plant was best, both for a clean-energy base and to ensure a smooth skill transfer for workers from the defunct coal-fired plant. In August 2017, the community and worker alliance declared victory when Canberra and the state government granted approval for the solar plant, which will become the biggest plant of its kind in the world. Work on the 150-megawatt facility will start in 2018, and it will be ready to dispatch energy to the grid by 2020. Port Augusta is an example of how one community, its workers, and their union can take the initiative and avoid the catastrophe that has descended upon other cities and towns built up around the fossil fuel industry. "There are too many communities to count who have sadly diminished or in some places died," Sharan said. "Often the impact on people beyond the workers can be devastating. We need to move rapidly to a zero-carbon future. The question is whether we can do that justly or unjustly."

∞

In the months following the closure of the Brunswick Mine, Ken Smith's confidence that he and his colleagues had done a "great job" in planning their own version of a just transition rapidly diminished. Although some of the mine's laid-off workers, such as Ken, found high-paying energy and mining jobs in cities and mining camps across Canada and beyond, many more were left behind, and unemployment rates in Bathurst soared. Lost without the security of the mine, scores of small businesses that had sold mechanical equipment, nuts, bolts, and truck parts to Xstrata went bankrupt and closed. "The mine was the focal point of that community," Ken says. "We were doing something new with the transition, but we still fell short. We failed to recognise that moving away was not an option for many of our brothers and sisters. We failed to understand how badly this would hurt our community."

While many of the mining communities in northern Canada rely predominantly on an itinerant workforce, the Brunswick Mine had employed locally, mostly fishermen left unemployed following the decline of the New Brunswick fishing industry in the 1960s. The familial and emotional ties to Bathurst of these third- and fourth-generation residents made it impossible for some to leave. "For some of our miners—because they had elderly parents with no other caregivers, or family members with special needs, or just a familiarity with the place where they had lived their entire lives—leaving was not an option," said Ken. "Now they are jobless, surviving on precarious work or employment insurance. Others are completely reliant on social assistance. These are people who worked for thirty or forty years in the mining industry, and this is now where they've found themselves."

For those like Ken who found work in the Alberta oil sands, or in mines farther north, the strain of a long-distance commute—travelling back and forth for thousands of kilometres to work exhausting three- or four-week shifts—has wreaked havoc on their personal lives. Although wages are sometimes three or four times the rate of those back home, many marriages have crumbled under the strain of separation. Ken counts himself lucky that his wife of more than thirty years made a last-minute decision to move with him to Fort McMurray, a difficult decision for the couple, as they were forced to leave behind his wife's beloved sister with special needs. Had his wife not accompanied him, Ken is adamant that he would not have lasted more than a couple of months on the oil sands. "It was very tough for me to leave Bathurst. I am fifty-six years old, and it is the first time I have been away from home. Believe it or not, even us old guys get homesick."

To cope with this newfound sense of displacement, Ken busied himself with union organising at Fort McMurray. During the day, he worked on the gargantuan dump trucks that lumbered throughout the oil sands facilities. The rest of the time, as president of the Unifor Local 707A, he worked to protect the interests of thirty-five hundred Suncor Energy oil sand workers. In

December 2015, that position led to a chance invitation to attend the UN climate summit in Paris as a union delegate. Sitting in the audience, listening to a panel discussion about clean-energy job creation, Ken felt uncomfortable that he and other fossil fuel workers were being demonised as climate change deniers, that oil sand workers such as him were just as toxic as the greenhouse gases that were choking up the atmosphere. "I have never been a huge environmentalist or anything like that," said Ken. "I'm just a guy who goes to work each day. But I also accept the science because people who have the expertise are saying that climate change is real. When I was a kid, the winters came earlier and the snow was deeper and stayed on the ground longer. There has been a gradual warming in my fifty years. I accept the facts."

Sensing an opportunity during the question-and-answer period, Ken stood up and took the microphone. He introduced himself as a fossil fuel worker. Many workers' attitudes in his industry had shifted, Ken told the hushed room, they "get it" that climate change is real. The trick, Ken said, was ensuring that workers such as himself—and their families—would not be left behind in the transition to clean energy. "[Fossil fuel workers] hope we're seeing the end of fossil fuels for the good of everybody," Ken told the room. "But how are we going to provide for our families? We're going to need some kind of transition. We've moved out there, we've invested in that industry—and when it ends, we're going to be left holding the bag."

Ken's brave and honest speech led to a rousing standing ovation, something that still makes him smile. "I have a grade-twelve education and am a little rough around the edges. I am always amazed when someone wants to hear my opinion." News of a fossil fuel worker calling for the demise of his own industry caused international headlines, catapulting Ken into the role of unlikely climate change hero. But to Ken, his speech at Paris was simply common sense: As someone who already knows what happens when your industry shuts down, he feels that he has something real to contribute to the discussion about what happens next. He just wishes that more people in the upper echelons of government and climate change policy would listen. "I was that miner with thirty-three years' service when the mine closed down; I was that union president who had to look at my members. Members who were my union brothers and sisters, my friends, my teammates, school friends—they were much more than just co-workers and they wanted to keep their jobs much more than they wanted 'transition.' Now I am again that union president in an industry that is the only game in town. In a remote community that is dependent on the oil sands for survival. Again, we are talking about transition, but this time the conversation goes a bit further to a just transition, and what that means. We need to look to our past and see what we did well and where we fell short. We have time to get this right, but we will only get one shot at it."

For Ken—and Sharan Burrow—getting it right means governments partnering with workers from sectors that are about to lose jobs. Fossil fuel workers will not resist change, Ken believes, as long as we take the fear out of the transition. Ken knows that consultation matters from the conversations that he had with his co-workers at Fort McMurray, people like him who came to the oil sands from other failed industries. "They all came to Fort McMurray because they didn't want to end up on welfare," Ken says. "They wanted to do better, they wanted their children to have opportunities, they wanted to provide. I know that these people will not resist change if they know that their families will be protected. That is how you make partners out of the workforce. Preparation is way better than resistance. We know that the tide is coming in. Let's prepare ourselves to move to the next job."

"He is a treasure," Sharan Burrow says of Ken Smith. It is people such as Ken whom Sharan kept in the forefront of her mind as she and other leaders fought tirelessly and succeeded in getting a reference to Just Transition into the Paris Agreement. Now Sharan is trying to implement a "just transition strategy" at the national and local levels to encourage countries to ensure that all parties—workers, unions, companies, local and state governments—work together as part of a sustainable, low-carbon economy and benefit from decent and green jobs. Currently the ITUC is working with a range of cities around the world—with Oslo and Sydney leading the way—and with several companies, including the B Team members, to move to zero carbon using a just transition strategy. Sharan hopes to meet a target of five cities and companies each year, a mammoth but achievable task that she believes is akin to the complexity and reach of the social contract known as the Marshall Plan, which successfully sparked the economic recovery of a decimated Europe between 1948 and 1951.

The ITUC's just transition strategy means finding jobs for affected workers who want them, providing training for those who need it, and ensuring decent pensions with health care for workers who cannot find new jobs. Ken is adamant that a just transition must be about more than providing social assistance. He wants to avoid the "knee-jerk" reaction of providing employment insurance—Canada's unemployment benefits—as a long-term strategy, something that he saw following the closure of the Brunswick Mine. "Employment insurance was never put in place to be a long-term safety net," Ken said. "It was a bridge to your next job. Employment insurance, for me, is like putting a Band-Aid on open-heart surgery. It doesn't stop the bleeding." Instead, Ken and Sharan argue, governments must drive investment to places where jobs are lost and stimulate economic development. "We should really be putting the heat on governments to force these companies who know that climate change is coming, and that they will be downsizing, to invest in other job creation. It's very simple," said Ken.

When Ken saves enough leave, he often returns to Bathurst with his wife to visit their daughter who still lives in the town, and to see his sister-in-law. Sometimes during these trips home he will slip out of his house and drive his truck out to the old Brunswick Mine site. Although the mine structure—the smelter, mineshafts, mill, and office buildings—have long been dismantled, the former Brunswick Mine still cuts a dramatic scar across the landscape. A deep tailings pond—a massive slurry hole used to store the water and waste from the mine—sits in the foreground of a large grassy hill that is sprayed every year to neutralise the toxic seepage that continues to bubble up from beneath the ground. But when Ken looks across this barren landscape, he sees only the bustle of the mine as it was in its heyday and smiles at the memories of friendships with other men forged amid blood, sweat, and grime. "We had fatalities, injuries, and strikes at the Brunswick Mine. But we try to remember the good times and the humour. I can stand there and look across and see everything, where the mill was, the crosses, where good things happened, and bad. There are a lot of memories there." Now, when Ken meets up with his old Brunswick mining pals, the talk inevitably turns nostalgic. "We say that we didn't know how good we had it, until we lost it. None of us have found another working situation that was as comfortable as we had. We worked with our friends and our family and we became a bigger family. That is what we have lost, but it is also what we remember."

Ken's new family is composed of the thirty-five hundred Suncor employees in Fort McMurray who are relying on their union president to help them with the inevitable transition away from fossil fuels. This time—with the help of Sharan Burrow and the ITUC—Ken is focused on an all-inclusive approach that will not just centre on the needs of his workers but will also protect the wider community. "I want to be the guy looking out for the worker, and for the community that has protected that worker. I can't allow thirty-five hundred workers and their families to be thrown out on the street. Failure is not an option; there is simply too much at stake."

To Christiana Figueres, a key architect of the Paris Agreement, one hopeful example of progress is her native Costa Rica, which now produces nearly all of its electricity through renewable energy. "Imagine that, my little country holding the world record!" she says. (Photo by Julien Paquin)

10

PARIS—THE CHALLENGE
OF IMPLEMENTING

A T PRECISELY 7:16 P.M. on December 12, 2015, in a conference centre at Le Bourget Airport outside Paris, the foreign minister of France, Laurent Fabius, picked up a green-topped gavel and brought it firmly down onto the table. In that instant, after two weeks of marathon negotiating efforts, the world's first global deal to help stave off the worst effects of climate change was born. The assembled delegates—world leaders and government ministers, diplomats, business leaders, and civil-society representatives—leapt to their feet and started clapping wildly. Then the cheering started, and the hugging. Looking around, I saw many of us were in tears.

After two decades of false starts and dead ends—and just six years after the acrimony of Copenhagen—such a deal had once seemed impossible. Now a new and fair accord—the Paris Agreement—set the course for a historic transformation of the world's fossil fuel economy by binding rich and poor countries alike to limit their emissions to safer levels. Poor nations would need to receive billions more dollars to help cope with the effects of extreme weather and to transition to a greener economy powered by renewable energy. In a city still reeling from a night of terror attacks a month earlier that had killed 130 people, the Paris Agreement was a triumph of hope over darkness: the best chance yet that the world could begin to avert the most devastating effects of our warming planet.

Having endured two weeks of little sleep, and long days spent in temporary buildings on the outskirts of Paris, we were overwhelmed with exhaustion, joy, and pride. The Paris Agreement was not just a historic turning point in a race to avert the potentially disastrous consequences of an overheated planet, but a resounding endorsement of the principles of climate justice. It was, as I would tell a reporter later that night, in every sense an "agreement for humanity." The architects of the agreement acknowledged the importance of climate justice in the text and made commitments on human rights and gender equality, agreed on a framework for monitoring national progress, and persuaded rich countries to provide financing for climate action in poorer nations. By committing to limit global warming to "well below" a 2°C rise above pre-industrial temperatures, the agreement also recognised the plight of Anote Tong and the people of Kiribati, along with the other forty-seven of the world's poorest countries, whose specific needs had, until this moment, been overlooked within the power politics of climate talks. Now Tong and other small-island state leaders such as Tony de Brum from the Marshall Islands could return home, their heads held high, and tell their people that their countries might still be saved. By agreeing to reduce greenhouse gas emissions to net zero by the second half of the century, the Paris Agreement acknowledged the work of Hindou, Constance, and Patricia. Now these women could return to Chad, Uganda, and Alaska and tell their communities that the world was moving away from fossil fuel dependency towards cleaner, sustainable forms of energy, land use, and waste management.

December 12 was already an auspicious date in the Robinson family calendar because forty-five years earlier, my husband, Nick, and I had exchanged marriage vows. Thirty-three years later, on the very same date, our first grandchild, Rory, was born. Exactly twelve years to the day since I had first peered into Rory's eyes and realised that the fight to limit climate change would be so central to my work, I was now celebrating a landmark global accord that might offer this young boy, and his generation, a chance to inhabit a better world.

∞

On the wall of her office, Christiana Figueres, the former head of the United Nations Framework Convention on Climate Change (UNFCCC) and a key architect of the Paris Agreement, keeps a framed motto: "Impossible is not a fact, it's an attitude." This aphorism allowed Christiana to achieve what many thought was unattainable in the wake of the Copenhagen talks: to assume the leadership of the UNFCCC and spend six gruelling years learning from the mistakes of Copenhagen to craft an agreement that would work for all countries—poor and rich alike. Recognising the failure of the 1997 Kyoto Protocol, where only developed countries undertook to reduce their greenhouse gas emissions, the office of French president François Hollande, in the lead-up to Paris, encouraged participating countries—no matter the size of their GDP—to submit a plan outlining how, and by how much, they would reduce carbon output. These plans—known as intended nationally determined contributions, or INDCs (another snappy acronym!)—became a historic break-through when more than 190 countries agreed to meet their targets. Participating countries will be required to reconvene every five years with updated plans that will further tighten their carbon output, and to report every five years, beginning in 2023, on their progress. Although there are no penalties for noncompliance, the hope is that countries will keep to their INDC targets primarily out of concern for the planet, and secondarily for the risk of being publicly shamed by their global peers as climate change footdraggers.

On June 1, 2017, the fears that had kept me awake that November night in the Marrakech guesthouse were finally realised when President Donald Trump announced that he was pulling the United States out of the Paris climate deal. From my home in Ireland, I watched on television as Trump, standing at a podium in the White House Rose Garden, denounced the Paris accord as a "draconian" deal and claimed that abandoning the agreement would reassert American sovereignty. I knew that America, the world's second-biggest polluter after China, was critical to the agreement's success, as its INDC pledge alone would account for more than a fifth of all emissions to be saved through 2030. In addition, America's failure to meet its commitments to the Green Climate Fund and other climate financing would make it even more difficult for developing countries to move to renewable energy. It is unconscionable that the United States has simply walked away from its responsibility to people both at home and abroad, in the interest of short-term fossil fuel profits, and abandoned an agreement that was negotiated by more than 190 world leaders, over decades, in the interests of all people and the planet.

But my fears that America's exit from Paris would undermine the deal did not materialise when two responses became clear. Firstly, the rest of the world planned to redouble their own climate commitments. Within minutes of Trump's announcement, the leaders of France, Germany, and Italy issued a joint statement saying that the agreement was "irreversible" and that their countries were staying the course. By the next day, editorials in leading newspapers and news on social media showed a global chorus of defiance and criticism against the decision. From China to Russia to India and across the European Union, successive world leaders clamoured to reassert their commitment to Paris. In a televised speech from Paris, France's recently elected president, Emmanuel Macron, invited American climate scientists to continue their work in his country and vowed to "make our planet great again." From Berlin, Chancellor Angela Merkel denounced Trump's action and said that it would "not deter all of us who feel obliged to protect this earth."

In the United States, an alliance of cities, states, and businesses—led by California governor Jerry Brown and former New York mayor Michael Bloomberg—committed to forging ahead with their individual state and city plans to lower emissions and to work separately with the United Nations Framework Convention on Climate Change. "We're going to do everything America would have done if it had stayed committed," Bloomberg said. By week's end, the message was clear: This was a crisis with no place for partisan rhetoric, and the world—with or without the government of the United States—would move forward in the fight to tackle climate change. As Christiana Figueres would observe in the late summer of 2017, Trump's decision to exit the Paris Agreement had mobilised the rest of the world, creating a huge groundswell of support for Paris that the climate movement could scarcely have achieved on its own. "I've already begun writing my thank-you letter to President Trump," Christiana observed wryly.

∞

In the past three years, global emissions of carbon dioxide from the burning of fossil fuels have levelled after rising for decades. More encouragingly, these emissions stayed flat while the global economy and the GDPs of major developed and developing nations grew.[1] This is good news, a promising indication that our work in climate mitigation is starting to pay off. But despite these positive signs, an unprecedented global effort is still required to hold warming to well below 2°C above pre-industrial levels and to save Kiribati and the lives of millions of vulnerable people along the world's coastlines. Even if all countries met their targets set out by the INDCs in the Paris Agreement, scientists predict that we would still experience a global temperature increase of more than 2.7°C.

We face a difficult truth: While Paris remains an unprecedented success, it is also a fragile foundation for action. The movement to address climate change—and to promote climate justice—must now shift to a new stage, with urgency and determination. All of us—governments, both powerful and small, prosperous and impoverished; cities, communities, business leaders, and individuals—bear responsibility. We must all take up this opportunity. The threat to our planet may be dire, but the potential opportunity is also historic—the chance to stop an existential threat, to conquer poverty and inequality, and to empower those who have been left behind and neglected.

As we pursue this new stage of bold action, we will succeed only if we recognise that the struggle to combat climate change is inextricably linked to tackling poverty, inequality, and exclusion. If we keep that link foremost in our minds, our solutions will be more effective and more enduring. Economic growth built on sustainable energy and land use will safeguard the lives of the most vulnerable from the effects of climate change and offer the best chance of lifting more communities out of poverty. If we give voice to those who have been marginalised and shut out, our policies and projects—both public and private—will tackle the root causes of both climate change and inequality. If we follow the example of those individuals on the front lines of climate change, we can find silver linings of resilience and hope in the belief that we can effect change. Such as Constance Okollet, who plants mango, avocado, and orange trees in her village in eastern Uganda to stop topsoil erosion and to prevent flooding. Or Natalie Isaacs, who brings her kitchen-table movement into homes across the world, inspiring women to change their lives in small ways that will make a big impact on our global carbon footprint. Or Sharon Hanshaw, the accidental activist, who used her voice to highlight the injustice that her marginalised community felt in the aftermath of Hurricane Katrina.

I often think of my father, a family doctor, whose life was transformed by the introduction of rural electrification across Ireland. I can still remember the awe in my father's voice as he described the revolution that the mere flick of a switch brought to his daily practice. Thanks to electric light, my father no longer had to deliver babies or tend to broken bones and wounds by candlelight. Electric pumps provided fresh water directly into his patients' homes, lightbulbs replaced dull and dangerous oil lamps, rural industry flourished, and the radio ended social isolation, bringing news and entertainment to rural families across the country. But the harsh reality is that as many people across the world today live without electricity as existed in the world when Thomas Edison first invented the lightbulb. Without reliable access to electricity, doctors cannot provide clinical services after sunset. Patients in the developing world cannot benefit from X-rays, ultrasounds, or incubators. Vaccines and medicines cannot be stored, and doctors cannot communicate with other health-care professionals. Nearly three billion people still live without access to clean cooking. Instead, to cook they rely on high-polluting solid fuels—wood, charcoal, animal dung, and crop waste—with fumes that kill more than four million people every year,[2] mostly women and children in Africa and Asia, and sicken millions more.

Providing electricity to the 1.3 billion people who lack access across the developing world remains one of the largest challenges on earth. Development is not possible without energy, but we must follow the goals set out in the Paris Agreement and create access to clean, affordable, and sustainable electricity. Inspiring examples of countries in the developing world spearheading solutions in renewable energy already exist. India, the third-largest emitter of carbon dioxide, where 240 million people still lack proper access to electricity, has the option to use coal to rapidly expand the country's electrical grid, but the Indian government has committed to providing electricity to all its people by 2030 by becoming a global leader in solar power. This includes an ambitious target to generate 160 gigawatts of wind and solar power by 2022. Thanks to $1 billion in support from the World Bank,[3] the Indian government will work to place rooftop solar panels on houses across the country, which will provide energy for Indian children to study at night and for families to refrigerate and cook their food.

In India's westernmost state of Gujarat, women cook with clean fuel and power up their cell phones using solar roof panels. Rachel Kyte, chief executive officer of Sustainable Energy for All, and a special representative of the UN secretary-general, says the traditional way of connecting people to the grid—via electrical poles, copper wiring, and cheap coal—no longer applies in this new age of solar and clean power. "The cheaper, faster, and easier way to provide people in the developing world with energy is with off-grid renewable systems," Rachel says. Once villages are electrified and have access to clean cooking, they will have access to better health care and schools with electric light where children can study for longer.

Empowering individuals without access to basic services is the goal of Sheela Patel, who works to provide water, sanitation, and electricity to the more than one billion people living in slums around the world. Sheela is chair of the Slum/Shack Dwellers International (SDI), a network of community-based organisa- tions of the urban poor in thirty-three countries and hundreds of cities and towns worldwide. Given the shoddy construction of buildings in slums and informal settlements, these areas tend to be hardest hit by extreme weather events and face extra urgency when it comes to climate resilience. In 2014, to help these communities better prepare for the inevitable onslaught of climate change, SDI launched the Know Your City Campaign to profile and map slum settlements and to use the data to upgrade the cities and manage climate risks. The data and mapping allows slum dwellers to "reblock" their towns by physically rearranging themselves to create new streets and public spaces that allow for the introduction of electricity and sanitation, and that provide each residence with an address. To date, SDI has mapped approximately five hundred cities and more than seven thousand slums. Across East, West, and South Africa, SDI has helped introduce twenty-one energy-service hubs across eight countries that now provide solar power to 15,000 slum households. Across the SDI network, the federations have extended clean water to approximately 185,000 households and built toilets for 220,000 more. By helping slum dwellers in Monrovia to remap their settlement, or village women in Gujarat to fix solar panels to their roofs, Rachel and Sheela demonstrate that many climate change solutions can be found in the developing world. We will all benefit if the peoples of the developing world are supported with incremental finance and greater access to technology, on a scale that the international community has often promised but has rarely managed to deliver. This isn't aid or charity. In the fight to tackle climate change, it is enlightened self-interest.

∞

In April 2017, a group of scientists, business leaders, and climate activists released a groundbreaking report that identified the year 2020 as a game-changing opportunity to turn the tide on global warming. The report warned that if emissions continue to rise beyond 2020 or even remain at their current level, the temperature goals enshrined in the Paris Agreement will be impossible to attain. Christiana Figueres has embraced this 2020 deadline and created a new initiative, Mission 2020, that aims to bend the greenhouse gas emissions curve downwards by 2020. It may be the only chance that we have. Christiana admits that the Paris

Agreement did not include an immediate plan of action to bring emissions to a rapid decline by that year. "Urgency is not embedded in the Paris Agreement," said Christiana. "We did put in a long-term target of full decarbonisation by the second half of the century, but when you walk that back, then you realise that you have to be at the climate turning point by 2020." The good news, Christiana believes, is that the Paris temperature goals can still be met if emissions begin to fall by 2020.

Bringing the best minds together, Christiana has devised a bold but achievable road map across six sectors—energy, transport, infrastructure, land use, industry, and finance—that business leaders, investors, and policymakers can follow. The road map showcases climate solutions that are already working, including plans for achieving 100 percent renewable electricity production while guaranteeing that markets can cushion renewable energy expansion.[4] Christiana points to her own country, Costa Rica, which in 2015 produced 99 percent of its electricity supply through entirely renewable energy. The small Central American country, which uses a mixture of renewable sources including hydro, wind, biomass, and solar, hopes to be entirely carbon neutral by 2021. "Imagine that, my little country holding the world record for electricity produced by renewable energy!" said Christiana. "It's a pretty impressive trajectory, and one that other countries can follow." Much farther south, Uruguay, which has invested heavily in wind and solar, now receives 95 percent of its electricity supply from renewable sources.

The Mission 2020 plan targets the Goliaths of fossil fuel consumption who are making promising commitments towards renewable energy dependency. China, the world's largest emitter of greenhouse emissions, has become a renewables superpower, solidifying a global lead in solar, wind, and other renewable sources. Although China still invests heavily in coal, in 2016 it added more than 34 gigawatts of solar capacity, more than doubling its solar capacity in that one year.[5] China already produces two thirds of the world's solar panels, and nearly half of the world's wind turbines. Under its current Five Year Plan, the Chinese government aims to have no less than 750 gigawatts of renewables capacity available by 2020,[6] more than that of all of the countries of the Organisation for Economic Co-operation and Development (OECD) combined. India, the world's third-largest emitter, recently revised its Paris Agreement targets to provide 60 percent of its electricity through renewable energy by 2027, three years sooner than it had stipulated. Globally, estimates suggest that solar power alone could supply nearly 30 percent of the world's electricity needs by 2050.[7] The launch in August 2017 of Al Gore's film *An Inconvenient Sequel*, and his powerful advocacy around it, helped to reinforce the momentum towards renewable energy and issues of sustainability.

But the stories of developing countries on the front line of climate change that are turning their backs on fossil fuels and transitioning to renewable energy inspire me the most. Although two thirds of its people currently have no access to electricity, Ethiopia has pledged to achieve ambitious emissions reductions and invest in renewable energy by 2025.[8] Fiji, which may one day house the nation of Kiribati, has committed to be fully dependent on renewables by 2030. In 2017, more than half of Fiji's energy production came from hydropower, an annual increase above 20 percent. Kenya, where only half of the people have access to electricity, has cut electricity import costs by 51 percent by harnessing geothermal power. Kenya intends to generate more than 70 percent of its electricity through renewable energy, including wind and geothermal power, by 2030.[9] In 2015, more than half of the world's renewable energy investment of $286 billion was for energy projects in developing or emerging countries.[10]

Critical to achieving the Paris Agreement goals are the city and state governments that can help meet the ambitions of Paris through transport and infrastructure policy. This subregional approach was strongly championed by Laurence Tubiana, France's ambassador for climate change leading up to Paris. In November 2016 at the UN climate conference in Marrakech, she and Hakima El Haite, as high-level climate champions, launched the 2050 Pathways Platform to support long-term low-greenhouse-gas-emissions development strategy. The governor of California, Jerry Brown, has become one of America's de facto leaders on climate change; dozens of states and some 160 cities across America have pledged to help the country's commitment to Paris with or without support from Washington. Since much of the work on reducing emissions happens at the state and local level—through renewable energy mandates set by utility commissions, efficiency rules for appliances, and fuel-mileage standards—this regional American alliance could well make a difference. California, a compelling climate case study, already shows how multiple solutions can work in tandem to lower carbon emissions and still rapidly expand an economy. In 2006, the state's visionary climate law pledged to reduce greenhouse emissions to 1990 levels by 2020, a goal that the state is on track to meet.[11]

The C40 Cities Climate Leadership Group, a network of ninety global cities committed to fighting climate change and including New York, London, Paris, Sydney, and Seoul, has adopted its own strategy called Deadline 2020, which aligns its carbon-lowering plan with the Paris Agreement. Around the world, cities and states have initiated action plans to fully decarbonise buildings and infrastructures by 2050.[12]

France's president too has become a prominent advocate for subregional action to combat climate change. On December 12, 2017, the second anniversary of the Paris Agreement, I was invited to attend the One Planet Summit convened by President Emmanuel Macron. On the afternoon before the summit, President Macron invited a group of the Elders[13]—including myself, Kofi Annan, Gro Brundtland, Lakhdar Brahimi, and Ban Ki-moon—to meet with him at the Élysée Palace. We discussed a range of problems, including the Middle East, the refugee crisis, North Korea, and Myanmar. President Macron outlined his rationale behind hosting the One Planet Summit,[14] expressing concern—a sentiment that we shared—that not enough progress had been made in the two years since the Paris Agreement to meet the goal of staying well below 2°C and working to 1.5°C. President Macron had invited heads of state and government, governors, city mayors, and business, philanthropic, and civil society leaders to attend the summit on the condition that they would be prepared to commit to raising their ambitions significantly, and at the summit twelve major coalitions for concerted action were announced. To ensure accountability, President Macron announced to the Elders that a second summit would be held in December 2018 to measure implementation of the commitments made. Listening to President Macron speak from across the table, I was suddenly struck by a thought: the current president of France was younger than my two older children!

I am closely involved with one example of business leadership called the B Team, which brings together a growing number of business leaders from around the world, including Sir Richard Branson of the Virgin Group and Jochen Zeitz of the Zeitz Foundation. Members of the B Team are committed to delivering a new way of doing business that prioritises people and the planet alongside profit—a "Plan B" for business. The B Team have been open to taking on new challenges—at a meeting in the margins of the World Economic Forum in January 2015, the B Team companies committed to reaching net-zero greenhouse gas emissions by 2050. This was an ambitious goal, and no other businesses were making such bold commitments. By listening to climate scientists and challenging each other, the B Team adopted this goal and used it as part of its advocacy in the lead-up to Paris. If these companies can demonstrate how they are moving to zero emissions while protecting workers via a just transition and respecting human rights, they will be powerful influencers in the business world.

If we are to meet the target of peaking emissions by 2020, we need more businesses to follow the example set by the B Team. We need a wholesale change in how business is done and money is invested—a little corporate social responsibility or greening will not be enough. Radical changes to supply chains, energy use, procurement, and even marketing will be needed if we are to keep warming below 1.5°C. Success stories are heartening, and it is crucial that they are shared. Publicising those countries, provincial governments, and businesses that are meeting—or even exceeding—their emission targets will inspire others and help raise the bar. So too will amplifying voices who carry moral authority, such as that of Pope

Francis, whose sweeping manifesto on the environment, *Laudato si'*, laments the devastation we have wrought on our dangerously degraded planet. In the encyclical, echoing the language of climate justice, Francis calls access to water "a basic and universal human right" and decries that "warming caused by huge consumption on the part of some rich countries has repercussions on the poorest areas of the world, especially Africa, where a rise in temperature, together with drought, has proved devastating for farming." It is not enough to save the planet, Francis argues; we must also curtail our obsession with consumption, which is destroying our Earth.

History has shown, time and time again, that when we come together, we are capable of big things. Our world today, despite its inequality and pain, is far better in so many ways than before. Collective action halved global illiteracy between 1970 and 2005. Worldwide life expectancy has risen from just forty-eight years in 1950 to over seventy-one today. In the last twenty-five years, child mortality around the world has halved. This is proof that we can succeed in tackling enormous developmental and existential challenges when we put people at the centre of everything that we seek to do.

∞

I think a lot about that world of 2050, when my six grandchildren will be in their thirties and forties and will share the world with more than nine billion people. How will they live together in social harmony, having enough food, water, and access to health, education, and overall well-being? We need a different way of living together, and it has to start now. It requires us to begin to sow the seeds of human solidarity and develop a global spirit of compassion. The existential threat of climate change has brought home our interconnectedness, and our dependence on one another, as never before, as the three powerful hurricanes that spun out of the Atlantic in 2017 have shown, ravaging rich and poor neighbourhoods alike across the Gulf Coast, Florida, and Puerto Rico. No one country can go it alone, and the issue is far too important to be left solely to politicians. At the same time, governments need to keep their commitments and indeed raise their ambitions to enable all other actions on climate change to be effective, to protect human rights, and to be inclusive and gender responsive.

What I have learned from those who inspired me to tell their stories is that we need to take personal responsibility for our families, our communities, and our ecosystems. And we need to do it with empathy and support for those less responsible for the climate problem, who are suffering more. They are showing the way despite having more barriers to overcome. The responsibility to implement the goals of the Paris Agreement must be taken to a lower level again. It has moved from just nation-states to now be shared by regions, cities, and businesses. The time has come to bring it home to families and communities. Each of us could work out our individual responsibility to live more sustainably. Better still, we could work on our families' responsibility and try to bring this to a community level. All schools, colleges, and workplaces could work out their responsibility to live more sustainably. Doing this with a conscious empathy for those who are most affected by climate change and least responsible would build the solidarity we need to ensure that developing countries can develop without emissions and lead to a fairer, more equal, more people-centred, more climate-just world.

In these moments, I think of Wangari Maathai, the Nobel Peace Prize laureate, environmentalist, and human rights activist. Wangari had a gift in recognising the interconnectedness of local and global problems, and in enlisting grassroots communities, particularly women, to create solutions. Just like Constance in Uganda, Wangari understood that her people could reverse deforestation and soil erosion in her native Kenya if they only planted trees, one by one. Ridiculed and threatened by both the Kenyan government and the country's powerful tree loggers, Wangari persisted. Now, the organisation that Wangari started in Kenya in 1977, the Green Belt Movement, has planted more than fifty-one million trees. Before her death in 2011, Wangari said, "In the course of history, there comes a time when humanity is called upon to switch to a new level of consciousness, to reach a higher moral ground."

We have arrived at that time. And we must reach for that higher ground.

ACKNOWLEDGEMENTS

T HIS BOOK BEGAN by accident. In January 2016, Nick and I were in New York and invited Lynn Franklin, my agent for my memoir, *Everybody Matters*, and George Gibson, the editor of the U.S. edition published by Bloomsbury, for a drink at our hotel to mark Lynn's retirement. During our conversation George asked how the work of my foundation on climate justice was going. I confessed my belief that the only way to convince people about the reality of climate change was to tell the stories of those affected—their courage and their resilience. Then George posed a challenge: "I would love you to write a storytelling book about climate justice. If you did, I would edit it and Bloomsbury would publish it." Lynn then joined in: "And I would retire from everything else, but stay on for this book!" We all laughed at this unexpected turn, but an underlying seriousness was clear.

My daughter Tessa had helped in writing my memoir, but was unavailable for this project because she had returned to full-time legal practice. Fortunately, Lynn proposed that I meet Caitríona Palmer, who had just published her own beautifully written memoir, *An Affair with My Mother*. It was an inspired idea, as Caitríona and I got on very well from the beginning, and our friendship deepened as the work progressed. This book became for all of us a true labour of love. We were on a mission to bring home the urgency of human-induced climate change, and that it had to be addressed through the lens of climate justice.

The stories are the essence and strength of the book, so Caitríona and I owe an enormous debt to each of those who gifted their personal story so that we might truly understand: Constance Okollet, Sharon Hanshaw, Patricia Cochran, Hindou Oumarou Ibrahim, Jannie Staffansson, Vu Thi Hien, Anote Tong, Natalie Isaacs, Ken Smith, Sharan Burrow, and Christiana Figueres.

Not all stories could be included in full, but others interviewed who helped shape the book were Agnes Leina, Kathy Jetnil-Kijiner, Thilmeezza Hussain, Pa Ousman, Rachel Kyte, and Sheela Patel.

Some friends also helped by advising on the choice of stories and the overall balance of the book, in particular Celine Clarke and Bride Rosney, who in addition made valuable comments on the final draft. It was important that the stories be true to climate science, and I was grateful to Jennifer McElwain, professor of botany in Trinity College Dublin and a board member of my foundation, for her advice and amendments in that regard. I would also like to record my thanks and appreciation to Barbara Sweetman for her patience and professionalism in typing up the various draft chapters shared between Caitríona, based in Washington, D.C., and myself in Dublin and in other locations where I was advocating for and working on issues of climate justice. It is also appropriate to record my appreciation to the Ireland Fund of Monaco for an award financing Caitríona's residency for several weeks at the Princess Grace Irish Library.

Both Caitríona and I valued immensely the wise counsel and detailed editing by George Gibson, and we were appreciative that when he parted from Bloomsbury, he was asked to continue as the book's editor. I am delighted that Bloomsbury is the publisher worldwide of *Climate Justice: Hope, Resilience, and the Fight for a Sustainable Future* and has believed in it from the beginning.

At Bloomsbury USA, Nancy Miller, associate publisher and editorial director, and Ben Hyman, senior editor, deserve special mention; as do Alexandra Pringle, editor in chief, and Emma Hopkin, managing director, of Bloomsbury UK.

I take full responsibility for any errors in the book, but I wanted to end with warmest thanks to two people who were there every step of the way. It was a joy to work with Caitríona Palmer, and I admire how skilfully she helped me tell the stories with such empathy and deep understanding. And once again I benefitted from the beady eye and red pencil exercised by my husband and great ally, Nick, whose personal support has been invaluable.

NOTES

PROLOGUE

1. The Mary Robinson Foundation—Climate Justice, www
.mrfcj.org.

1. UNDERSTANDING CLIMATE JUSTICE

1. Brian Kahn, "This Graphic Puts Global Warming in Full
Perspective," Climate Central, April 19, 2017, www.climate
central.org/news/628-months-since-the-world-had-cool
-month-21365.
2. David Wallace-Wells, "The Uninhabitable Earth," *New York*,
July 9, 2017, nymag.com/daily/intelligencer/2017/07/climate
-change-earth-too-hot-for-humans.html.

2. LEARNING FROM LIVED EXPERIENCE

1. It was an El Niño year.
2. Karla D. Maass Wolfenson, "Coping with the Food and
Agriculture Challenge: Smallholders' Agenda," Natural
Resources Management and Environment Department,
Food and Agriculture Organization of the United Nations,

prepared April 2013, revised July 2013, www.fao.org/file
admin/templates/nr/sustainability_pathways/docs/Coping
_with_food_and_agriculture_challenge__Smallholder_s
_agenda_Final.pdf.

3. THE ACCIDENTAL ACTIVIST

1. Reilly Morse, "Environmental Justice Through the Eye of
   Hurricane Katrina," *Focus*, May/June 2008, 7–9.
2. Ibid.
3. Rachel Morello-Frosch, Manuel Pastor, James Sadd, and
   Seth B. Shonkoff, "The Climate Gap: Inequalities in How
   Climate Change Hurts Americans & How to Close the
   Gap," dornsife.usc.edu/assets/sites/242/docs/The_Climate
   _Gap_Full_Report_FINAL.pdf.
4. Leslie Eaton, "In Mississippi, Poor Lag in Hurricane Aid,"
   New York Times, November 16, 2007, www.nytimes.com
   /2007/11/16/us/16mississippi.html.

4. VANISHING LANGUAGE, VANISHING LANDS

1. April M. Melvin et al., "Climate Change Damages to
   Alaska Public Infrastructure and the Economics of Proac-
   tive Adaptation," *Proceedings of the National Academy of Sciences
   of the United States of America* 114, no. 2 (2016): E122–E131,
   www.pnas.org/content/114/2/E122.abstract.
2. Arnoldo Valle-Levinson et al., "Spatial and Temporal Vari-
   ability of Sea Level Rise Hot Spots over the Eastern United
   States," *Geophysical Research Letters* 44, no. 15 (August 2017):

7876–82, onlinelibrary.wiley.com/doi/10.1002/2017GL073
926/abstract.

6. SMALL STEPS TOWARDS EQUALITY

1. Mike Ives, "In War-Scarred Landscape, Vietnam Replants
    Its Forests," *Yale Environment 360*, November 4, 2010, e360
    .yale.edu/features/in_war-scarred_landscape_vietnam_re
    plants_its_forests.
2. "The Context of REDD+ in Vietnam: Drivers, Agents and
    Institutions," Center for International Forestry Research,
    2017, www.cifor.org/library/3737/the-context-of-redd-in
    -vietnam-drivers-agents-and-institutions/.
3. Ibid.
4. Ibid.
5. Ibid.
6. Don J. Melnick, Mary C. Pearl, and James Warfield,
    "A Carbon Market Offset for Trees," *New York Times*,
    January 19, 2015, www.nytimes.com/2015/01/20/opinion
    /a-carbon-offset-market-for-trees.html?mcubz=1.
7. *Stern Review: The Economics of Climate Change*, http://unions
    forenergydemocracy.org/wp-content/uploads/2015/08
    /sternreview_report_complete.pdf.
8. Equivalent to 54.546 billion metric tons of carbon.
9. "Toward a Global Baseline of Carbon Storage in Collective
    Lands," Rights and Resources Initiative, November 2016,
    www.rightsandresources.org/wp-content/uploads/2016/10
    /Toward-a-Global-Baseline-of-Carbon-Storage-in
    -Collective-Lands-November-2016-RRI-WHRC-WRI
    -report.pdf.

10. Ibid.

11. "Context of REDD+ in Vietnam."

12. Indigenous peoples and local communities customarily claim at least 50 percent of the world's lands, but legally own just 10 percent.

8. TAKING RESPONSIBILITY

1. "State of the Climate 2016," Australian Government Bureau of Meteorology, 2017, www.bom.gov.au/state-of-the-climate/.

2. Damien Cave and Justin Gillis, "Large Sections of Australia's Great Reef Are Now Dead, Scientists Find," *New York Times*, March 15, 2017, www.nytimes.com/2017/03/15/science/great-barrier-reef-coral-climate-change-dieoff.html?mcubz=1.

9. LEAVING NO ONE BEHIND

1. Robert Jones, "Brunswick Mine Closes Bathurst-Area Operation," CBC News, May 1, 2013, www.cbc.ca/news/canada/new-brunswick/brunswick-mine-closes-bathurst-area-operation-1.1335287.

2. Nadja Popovich, "Today's Energy Jobs Are in Solar, Not Coal," *New York Times*, April 25, 2017, www.nytimes.com/interactive/2017/04/25/climate/todays-energy-jobs-are-in-solar-not-coal.html?mcubz=1&_r=0.

3. Anmar Frangoul, "9.8 Million People Employed by Renewable Energy, According to New Report," CNBC, May 24, 2017, www.cnbc.com/2017/05/24/9-point-8-million-people-employed-by-renewable-energy-according-to-new-report.html.

4. "2017 U.S. Energy and Employment Report," U.S. Department of Energy, January 2017, www.energy.gov/downloads/2017-us-energy-and-employment-report.

5. "Renewable Energy and Jobs—Annual Review 2017," International Renewable Energy Agency, May 2017, www.irena.org/menu/index.aspx?mnu=Subcat&PriMenuID=36&CatID=141&SubcatID=3852.

6. "Richard Branson's Big Idea for Building a Better Version of Capitalism," *Economist*, October 6, 2012, www.economist.com/node/21564197.

7. Mission 2020, www.mission2020.global.

8. Repower Port Augusta, www.repowerportaugusta.org.

10. PARIS—THE CHALLENGE OF IMPLEMENTING

1. Christiana Figueres et al., "Three Years to Safeguard Our Climate," *Nature*, June 28, 2017, www.nature.com/news/three-years-to-safeguard-our-climate-1.22201#/b1.

2. World Health Organization, "Household Air Pollution and Health," fact sheet no. 292, February 2016, www.who.int/mediacentre/factsheets/fs292/en/.

3. World Bank, "Solar Powers India's Clean Energy Revolution," www.worldbank.org/en/news/immersive-story/2017/06/29/solar-powers-india-s-clean-energy-revolution.

4. Figueres et al., "Three Years to Safeguard Our Climate."

5. Steve Hanley, "China Doubled Its Solar Capacity in 2016," *CleanTechnica*, February 9, 2017, www.cleantechnica.com/2017/02/09/china-doubled-solar-capacity-2016/.

6. Brian Wang, "Solar Power in 2020 World Will Nearly Triple Current Levels to about 450 GW and Global Wind

Power Will Be about 750 GW," *NextBigFuture*, March 22, 2016, www.nextbigfuture.com/2016/03/solar-power-in -2020-world-will-nearly.html.

7. Figueres et al., "Three Years to Safeguard Our Climate."

8. Federal Democratic Republic of Ethiopia, "Intended Nationally Determined Contribution (INDC) of the Federal Democratic Republic of Ethiopia," United Nations Framework Convention on Climate Change, www4.unfccc.int /submissions/INDC/Published%20Documents/Ethiopia /1/INDC-Ethiopia-100615.pdf.

9. Maina Waruru, "Kenya on Track to More Than Double Geothermal Power Production," *Renewable Energy World*, June 15, 2016, www.renewableenergyworld.com/articles /2016/06/kenya-on-track-to-more-than-double-geother mal-power-production.html.

10. Figueres et al., "Three Years to Safeguard Our Climate."

11. Paul Hawken, ed., *Drawdown: The Most Comprehensive Plan Ever Proposed to Reverse Global Warming* (New York: Penguin Books, 2017).

12. Figueres et al., "Three Years to Safeguard Our Climate."

13. An independent group of global leaders founded in 2007 by Nelson Mandela to work together for peace and human rights.

14. One Planet Summit, www.oneplanetsummit.fr.

INDEX

A NOTE ON THE AUTHOR

M ARY ROBINSON SERVED as the seventh, and first
female, president of Ireland, from 1990 to 1997, and as
United Nations High Commissioner for Human Rights, from
1997 to 2002. Robinson was honorary president of Oxfam
International from 2002 to 2012 and has chaired numerous
bodies, including the GAVI Alliance, vaccinating children
worldwide, and the Council of Women World Leaders (of which
she was a co-founder). She is a member of the Elders, an independent group of global leaders brought together by Nelson
Mandela. A member of the American Philosophical Society, she
is the recipient of numerous awards and honours, including the
U.S. Presidential Medal of Freedom and the Indira Gandhi and
Sydney Peace Prizes. She is president of the Mary Robinson
Foundation—Climate Justice and lives with her husband, Nick
Robinson, in Dublin.